The Commission of the European Communities

CIVIL ENGINEERING DESIGN FOR DECOMMISSIONING OF NUCLEAR INSTALLATIONS

This Report was prepared as part of the European Atomic Energy Community's cost sharing research programme on "Decommissioning of Nuclear Power Plants", Contract No. DE-G-002-UK.

CIVIL ENGINEERING DESIGN FOR DECOMMISSIONING OF NUCLEAR INSTALLATIONS

A A Paton, P Benwell, T F Irwin and I Hunter
(Taylor Woodrow Construction Limited, Southall, UK)

Published by Graham & Trotman Ltd. for the Commission of the European Communities

Published in 1984 by
Graham & Trotman Ltd., Sterling House, 66 Wilton Road,
London SW1V 1DE, UK.
for the
Commission of the European Communities, Directorate-General
Information Market and Innovation, Luxembourg.

EUR 9399 EN

ISBN-13:978-0-86010-614-2 eISBN-13:978-94-009-5632-2
DOI: 10.1007/978-94-009-5632-2

CONTENTS

1. INTRODUCTION

1.1 This report describes the work carried out by Taylor Woodrow Construction Limited (TWC) in a study aimed at identifying features which may be incorporated at the design stage of future nuclear power plants to facilitate their eventual decommissioning and, in so doing, promote economic and radiological benefits at the decommissioning stage.

1.2 For the purposes of this study, decommissioning of a nuclear facility means those measures taken at the end of the facility's operating life to remove it from the site and restore the site to green field conditions, and, while so doing, ensure the continued protection of the public from any residual radioactivity or other potential hazards present in or emanating from the facility. The overall decommissioning process involves eventual dismantling and demolition and may also include, where possible and appropriate, the intermediate steps of renewal and refurbishing.

1.3 The work has been carried out within an overall research programme on decommissioning organised through and partly funded by the Commission of the European Communities [1]. In addition to TWC itself, other contributions to the funding and technical aspects of the work have been made by the Nuclear Installations Inspectorate (NII), the Central Electricity Generating Board (CEGB), the National Nuclear Corporation Limited (NNC) and the Department of the Environment (DoE).

1.4 The initial brief for the study was that it should consider only civil engineering aspects of nuclear power plants based on gas cooled reactor systems. The brief, however, was extended to include an assessment of the civil engineering aspects of decommissioning light water cooled reactor systems, particularly the pressurised water reactor.

1.5 The work has been carried out in a number of sequential stages consisting principally of a literature review, identification of problems likely to arise in decommissioning, generation of possible solutions to the problems, first assessment of the feasibility of these solutions, closer investigation of promising solutions and, finally, preparation of conclusions and recommendations.

2. CONCLUSIONS AND RECOMMENDATIONS

2.1 Introduction

2.1.1 There is a substantial amount of literature available on decommissioning of nuclear power plants and, although much of it is repetitive and does little more than reflect on the subject, two key factors emerge. Firstly, it seems clear that no nuclear plant built to date has been designed with decommissioning as an important consideration. Secondly, concensus of current opinion appears to be that decommissioning of existing plants will be carried out in three stages, the last of these perhaps being delayed until approximately 100 years after final shutdown. At that time activity within the reactor cavity will have decayed to a level below which further decay is very slow and to which man may not be exposed for more than very limited periods. This would necessitate surveillance of each individual site for periods of at least 100 years after shutdown and raises the philosophical issue of whether future generations would be prepared to accept this situation with respect to public safety and protection of the environment.

2.1.2 The rapid advance of technology over the past 50 years has brought about the development of the nuclear reactor for commercial use in power generation and it is not unreasonable to anticipate that, in the next 50 years, continued development will lead to improvements in the methods used to handle and dispose of the active material created in nuclear installations.

2.1.3 For the present however, the principal technical problems to be overcome in dismantling gas cooled reactor plants lie in the demolition and removal of the prestressed concrete reactor vessel (PCRV) which is a dominant feature of current generation gas cooled stations and the biological shield, where present. In particular, these problems are associated with safe removal and disposal of the fuel, graphite and other internal components, followed by demolition and removal of the activated region of material of the PCRV structure immediately surrounding the reactor cavity. It is the latter of these activities on which the study has concentrated and, for the purposes of the work, the activated region has been taken as a layer of up to a maximum of 1m thick, including the liner, surrounding the cavity. In current PCRV designs, the steel in this region contains a variety of trace elements, notably cobalt, which, when subjected to neutron bombardment convert to radioactive isotopes which emit harmful gamma radiation for long periods after reactor shutdown.

2.1.4 For light water cooled reactor plants, dismantling problems are associated mainly with the reactor pressure vessel (RPV) and the primary shield wall which surrounds it. As far as the civil engineer is concerned, it is demolition of the latter that poses the major difficulties with respect to decommissioning. After shutdown of the plant, this wall will have an activated region approximately 1m to 1.5m thick, containing steel embedments and this will present problems very similar to those of the activated region of a PCRV.

2.1.5 It has not been possible within the scope of this study to consider either the commercial aspects of introducing design features to aid decommissioning or those aspects of deferring final decommissioning until long after shutdown. Clearly, one factor to be considered is whether the likely future availability of building land will be such as to make early re-use of the site a sufficient incentive to introduce features that will accelerate the decommissioning process.

2.1.6 The conclusions that follow in Sections 2.2 and 2.3 identify and comment on a number of features that could be considered at the design stage of future nuclear plants to eliminate or minimise the technical problems associated with dismantling that will arise when existing plants are decommissioned. Where further development of these features is considered worthwhile, this is stated, and a summary of the recommendations is given in Section 2.4. It should be recognised that these conclusions and recommendations represent ideas that the civil engineer may be able to offer to aid the decommissioning of a structure whose design, construction and operation encompass many different engineering disciplines. Any further development of these ideas would therefore need to be carried out in conjunction with scientists and engineers from these disciplines. It must further be recognised that a nuclear reactor system is expected to be durable and to be able to resist all predictable accidents and events that could in any way damage it or make it unsafe. It is therefore crucial that any measure introduced to make its demolition easier should not conflict with fundamental safety requirements.

2.1.7 The conclusions made are presented under a series of particular headings which follow the format in which the work of the study has been undertaken.

2.2 Gas Cooled Reactor Systems

2.2.1 Core Shielding

Since it is the presence of active steel in and attached to the concrete immediately surrounding the reactor cavity of a PCRV that largely creates the radiological problems which make demolition difficult, the most effective way of removing these problems would seem to be to provide a comprehensive and more efficient neutron shield around the reactor core. It is believed that provision of such a shield represents only a little more development of the core shielding arrangement designed for the Heysham II and Torness reactor vessels. For these vessels, calculations by reactor engineers have indicated that the activated region surrounding the cavity will be approximately 50mm thick. This feature is discussed in Section 6.2 of the report.

It is recommended that further work be done by reactor engineers to determine whether such a shield is feasible and to assess its effect on vessel size, radioactive inventory, permissible man access times following removal of the vessel internals, and reactor operational characteristics.

2.2.2 Planes of Weakness

Subject to compliance with operating and licensing requirements, one way of making the activated zone of a PCRV easier to remove would be to incorporate planes of weakness through it vertically, radially and in the hoop direction, thereby dividing the otherwise homogenous concrete into discrete, regularly shaped handleable pieces. Figure 11 shows the proposed layout of such planes for an assumed 1m thick activated region although, in practice, it would not be intended to extend the planes over the liner roof. The likely feasibility, from design considerations, of incorporating such a feature has been confirmed by preliminary analyses of two main service load conditions, namely, initial prestress and prestress plus proof pressure, and ultimate load. A simple stability check for seismic loading was also made.

The planes of weakness could be formed by constructing the activated zone either by means of precast blocks or by in-situ casting, each method having advantages and disadvantages. In practice it would probably be necessary to provide a thin layer of continuous in-situ concrete backing to the liner to provide anchorage and stiffening to it. It might similarly be necessary to cast a preformed concrete collar around penetrations where they connect with the liner. However, this concrete collar would be debonded from the planes of weakness zone and should not significantly affect the overall concept.

This feature is discussed in Section 7 of the report and it is recommended that further work should be done to confirm more positively the feasibility of the concept. This should include analyses of a PCRV with planes of weakness for the range of load combinations applicable during construction, commissioning and operation, and a more detailed assessment of how best to form the planes and of the materials to use. If this work confirmed present conclusions, consideration should then be given to supporting the analytical work by model testing.

2.2.3 Forceful Break-Up

Several possible methods of forcefully breaking up the activated region of the PCRV have been considered and some of these appear feasible. See Section 6.4 of the report.

Traditional explosive techniques appear to be a very effective way of breaking up homogenous concrete and a considerable amount of work has been done by TWC under Project 3 of the CEC programme on decommissioning [2,3,4]. This technique does, of course, have the disadvantage of noise, vibration, dust, flying debris and the like, depending on the size of the charges used. Where necessary, these disadvantages could be avoided by using, instead of explosives, chemical demolition agents, such as "BRISTAR", for unreinforced concrete, or mechanical devices such as hydraulic bursters.

Use of any of the above three techniques would be most effective if, during construction of the PCRV, suitable preformed holes were provided in the activated zone. Preformed holes, however, give rise to two possible problems. Firstly, in current PCRV construction, it is not permissible to leave, in the structure, any void that could form a shine path for neutron flux from the core. Preformed holes, therefore, would need to be plugged with a material capable of maintaining the shielding property of the concrete, able to survive immediately behind the liner during reactor operation yet able to be easily removed when required. Lead or graphite are two potential filler materials. Secondly, a large number of preformed holes immediately behind the liner may raise doubts on liner anchorage integrity and liner support. Further investigation of this idea could best be carried out against known requirements in a detailed design phase.

High temperatures, up to 900°C, can be used to degrade concrete. Whilst such a technique would be impractical on a large scale, it could be very useful for tackling localised problems.

A dominant problem in demolishing the activated zone of a PCRV is removal of the continuous, firmly anchored liner. One method of easing this problem may be to fill selected liner cooling pipes with liquid explosive which, when detonated, will split the liner plate into panels in readiness for demolition and removal of the backing concrete.

2.2.4 Selective Use and Location of Materials

It is believed that there are several options available for minimising the content in the activated concrete of materials, particularly steel, which are difficult to remove either for physical reasons or because of radiological effects. There may also be ways of altering the concrete mix to enhance its shielding properties. These aspects are discussed in detail in Section 6.5 of the report.

The development and use of steels with low content of trace elements such as cobalt, silver and niobium would have known benefits in reducing the radiation problem and work on this aspect has been carried out by Boothby and Williams.[5]

It is considered feasible to eliminate reinforcement and prestressing tendons from the activated zone. Whether or not this option would be adopted in future designs is a matter to be considered at the design stage, taking into account possible effects on design, plant layout, construction and overall cost.

Possible methods for reducing the amount of steelwork attached to the concrete side of the liner are considered in Section 2.2.5 of the report.

There are two ways in which the neutron shielding properties of concrete might be improved thereby leading to a possible reduction in the thickness and overall volume of activated material. The first is by increasing the hydrogen content of the mix and this could be achieved either by increasing the water content of the mix or by using hydrous aggregates.

The second, which also reduces the level of activity of the concrete, is by the introduction of boron, in the form of non-soluble boron frits, to the concrete mix. Prior to the adoption of these ideas, testing would of course be required to confirm that they gave the anticipated enhanced shielding properties, to establish procedures for producing the concrete and to check that neither caused any unacceptable effects on concrete workability, heat of hydration, creep, shrinkage, strength and durability.

It is recommended that a suitable programme to carry out the above testing be commissioned. The effects on gamma shielding and volume of waste should also be assessed.

2.2.5 Removal of Liner and Penetrations

One of the most important conclusions relating to liners is that liner designers see no alternative to mild steel liners firmly anchored to concrete, as in present PCRV designs. Hence, features that will aid removal of liners fall into two main categories, namely, those that may be considered for existing liner designs and those that envisage major changes in concept and approach to liner design. The idea of using liquid explosives in selected cooling pipes to split the liner into panels has been outlined in Section 2.2.3 above, and the full range of ideas considered for liner removal is discussed in detail in Section 6.6 and 8 of the report.

Some significant problems that will be encountered in removing existing liners arise from the presence of large fabricated steel anchors attached to them. In many cases, these became necessary because of late changes in design and space requirements. Elimination of such factors would obviously be beneficial in future designs.

Alteration to the detail of the connection between the liner and penetrations, as shown in figure 17, would enable disconnection of the penetration from the liner prior to separate, and easier, removal of both.

In so far as conceptual changes to liner design are concerned, it is believed that a significant simplification of liner removal could be achieved if the liner were largely detached from the vessel concrete. This would presuppose the deletion of the many stud or hook anchors which exist on present liners and whose essential purpose is to prevent the liner from buckling during the application of prestress to the vessel and as a result of operational temperature loading. It is considered possible that, if the liner could be isolated from the prestress loading and as much prestress creep as possible, the residual loading on it might be small enough to allow a significant reduction in the number of stud or hook anchors required, ideally leaving the liner attached to the concrete only at penetration locations.

Two ideas considered for achieving such an arrangement are a freestanding liner and a grouted liner. See Section 8.3 of the report. The freestanding liner concept is not a new idea[6] and essentially envisages a liner permanently separated from the concrete except at penetration locations. In order to prevent liner rupture under pressure loading the gap between the liner and the concrete would always need to be pressurised to reactor circuit pressure and the concrete surface would therefore need to be sealed, for example, with an epoxy coating. The grouted liner concept envisages the gap between the liner and the concrete being filled with grout at as late a stage as possible in the construction programme, certainly after prestressing and after as much as possible of the prestress creep had taken place. An outline of the grouted liner is shown in figure 39, with proposed penetration connection details shown in figure 40.

Preliminary work on the grouted liner concept has identified the likely main problems, typically, construction methods, connecting up cooling pipes and standpipes, assessment of degree of liner anchorage required and so on. It is recommended that further work on this concept should be carried out to determine whether or not it is feasible. It is also recommended that the previous work on the freestanding liner concept be followed up.

Closely linked to liner design, performance and removal problems is the liner insulation system which, in present PCRV designs, is a very intricate assembly. It is felt that a review of circulator design and gas circulation characteristics, aimed at reducing noise and vibration levels, might enable simplification of insulation system details. Further simplification might be possible if the concrete, or grout, immediately behind the liner could be designed to have enhanced insulating capability.

With regard to removal of the standpipes, some preliminary work has suggested that it may be possible to develop a scheme which allows stage by stage removal of the standpipe region, leading to shielded access to and removal of the reactor core and internals. This scheme is described in Section 8.5 of the report and outlined in figure 41. It is recommended that further work be done to assess the feasibility of this scheme and that of constructing all or part of the standpipe region as a separate removable structural element.

2.3 Light Water Cooled Reactor Systems

As far as the civil engineer is concerned, the only significantly activated region in a PWR plant is an inner 1.0m to 1.5m thick zone of the primary shield wall that surrounds the reactor vessel. This wall contains steel embedments and its decommissioning problems and their potential solutions are likely to be similar to those of the activated zone of a PCRV. Subject to structural and licensing requirements, one possible design feature particular to the PWR primary shield wall would be to construct the inner portion of the wall to contain a series of water tanks, as shown in figures 53 and 54. Such a feature would both reduce the amount of activated material in the wall itself and might also reduce the levels of activity in the remaining structures. It is recommended that further work be done to assess the feasibility of this idea.

2.4 Summary of Recommendations

2.4.1 In so far as gas cooled reactor systems are concerned, the detailed recommendations in respect of further work are as follows:

(a) Assess the feasibility and consequential effects of providing a core shield which, ideally, totally contains the neutrons emitted by the core.

(b) Further investigate the feasibility, from design and construction aspects, of providing planes of weakness in the activated zone of a PCRV. This work should include further detailed analyses, assessment of construction techniques and materials and, subject to the conclusions at that stage, supportive model test work should be considered.

(c) Initiate a programme of tests to investigate methods, and consequent effects, of improving the shielding properties of concrete, for example, by increasing its hydrogen content or by introducing boron as a constituent of the mix.

(d) Examine practical means by which a large number of preformed holes can be formed in the concrete immediately behind the liner without detriment to either the shielding properties of the concrete or the integrity of liner anchorage and support.

(e) Liaise with the working groups engaged on Project 3 with a view to facilitating the dismantling techniques they have developed.

(f) Extend the assessment of feasibility of a grouted liner concept for PCRVs, and follow up previous work on the freestanding liner concept.

(g) Investigate the development and use of a concrete or grout which, in addition to having the properties currently required for PCRVs, also has enhanced thermal insulating properties.

(h) Review circulator design and gas circulation characteristics with a view to reducing noise and vibration levels to give consequent simplification of insulation fixings.

(i) Develop and assess the feasibility of the integrated scheme for standpipe region removal leading to access to and removal of the reactor core and other internal components within the reactor cavity.

2.4.2 As far as light water cooled reator systems are concerned the recommendations made in 2.4.1 (a), (b), (c), (d) and (e) apply in principle with respect to demolition and removal of the primary shield wall. In addition it is recommended that work be done to assess the feasibility of constructing the inner part of the primary shield wall to contain water tanks.

2.4.3 For both reactor systems there are three simple measures which involve no great design innovation but which could significantly aid the decommissioning process and it is recommended that these measures be adopted. These are to ensure that as many as possible exposed surfaces are specified to be decontaminable, that as much as possible of the plant and buildings are easily demountable and that specific provision is made for removal of plant at the decommissioning stage. These measures would be part of an overall decommissioning plan which should be formulated before any detailed design of a nuclear installation is commenced.

3. LITERATURE REVIEW

3.1 The literature search and review consisted of the following sequential activities:

(a) Identification and collection of literature available either "in-house" or through "in-house" contacts.

(b) Identification of relevant references through a computer aided literature search of data bases N.T.I.S., COMPENDEX and PASCAL.

(c) Identification of additional references through miscellaneous other sources.

(d) Selective procurement, in hard copy, of a number of references identified in (b) and (c) above. The basis of selection was directed towards providing a representative sample of available references.

(e) From a brief examination of all the references obtained as described in (d) above, selection and further detailed study of those few which, when viewed collectively, were judged to contain all the basic information relating to the problems of decommissioning.

3.2 The main comment emerging from the literature review was that many of the references cover very similar ground and the actual information available on decommissioning experience is very limited. This situation is not altogether surprising when the relative infancy of nuclear power in commercial terms is considered. As existing nuclear power plants are progressively taken out of service it is to be expected that a store of knowledge and experience will be built up and begin to indicate design features that will aid decommissioning.

3.3 In the meantime, however, the literature pointed to several features whose consideration for nuclear power plants being designed now, or in the immediate future, could aid their decommissioning and dismantling.

3.4 Regarding type of plant, the literature review indicated that within the short to medium term future, nuclear power plants which will need to be decommissioned are based on the following reactor systems:

(a) Early magnox reactor, in steel pressure vessel with separate biological shield and secondary containment structure.

(b) Later magnox reactor, in prestressed concrete pressure vessel which acts as the primary and secondary containment, and as the biological shield.

(c) Advanced gas cooled reactor in multicavity prestressed concrete pressure vessel.

(d) Advanced gas cooled reactor in single cavity prestressed concrete pressure vessel.

(e) Light water cooled reactor in steel pressure vessel with separate biological shield and secondary containment structure.

3.5 The literature suggested that it is unlikely that existing, or near complete, stations based on any of the above reactor systems have in-built aids to the decommissioning and demolition of their civil works. Furthermore, of the above reactor systems, the concensus of opinion expressed was that the majority of future stations will be based either on the advanced gas cooled reactor with single cavity prestressed concrete pressure vessel or on the light water cooled reactor system. It was considered possible, however, that, in the long term, other reactor types such as the fast breeder reactor and the high temperature reactor, might be developed further and that these would be based on the multicavity style of prestressed concrete pressure vessel. It was believed that these systems would have similar decommissioning problems to the advanced gas cooled systems on which the literature survey therefore concentrated.

3.6 As to the decommissioning process itself, whilst many of the references studied do little more than philosophise on the various options that might be considered, it would now seem to be internationally accepted that decommissioning will be carried out in three stages, as follows:

(a) Stage 1 - This involves shutting the plant down, depressurising it, removing the fuel, making it saf and keeping it under surveillance. For a typical gas cooled reactor in the U.K. defuelling may take up to four years to complete.

(b) Stage 2 - This involves dismantling all plant and buildings external to the reactor biological shield and keeping what remains under surveillance.

(c) Stage 3 - This involves dismantling and removing the remaining plant and structures so that the site can be restored to green field conditions for unrestricted future use.

3.7 In general, Stage 2 would follow Stage 1, although if there were sufficient incentive to clear the major area of the site as early as practicable some of the Stage 2 work would be undertaken in parallel with Stage 1. Essential services would need to be retained and safety requirements satisfied. Likewise, it might be possible and desirable to commence Stage 3 in advance of completing Stage 2, again subject to provision for essential mechanical services and structural and radiological safety requirements.

3.8 Some scenarios postulated in the literature suggest that Stage 3 of the decommissioning process may be delayed until up to 100 years after shutdown of the plant.

4. REFURBISHING

4.1 As far as it has been possible to ascertain, the design specifications for nuclear power plants have not hitherto included any requirements relating to refurbishment of the plant. Generally speaking, items of mechanical and electrical equipment with limited service life or which require periodic replacement, either in whole or in part, are, of necessity, designed to facilitate this. It is also possible to envisage that, with varying degrees of difficulty, most items of plant in a nuclear power station could be dismantled, removed and replaced should it be considered economical to undertake either major renewal of the system or incorporation of improved equipment to extend the operating life of the station beyond its initial design life.

4.2 The difficulties of refurbishing arise primarily in those areas where very heavy large items have been embodied within structures and whose removal would entail major work on the fabric of the station. Particular problems exist in respect of the reactor system where dismantling and refurbishing may be prevented or inhibited by high levels of radiation. Such problems are most intense in relation to the pressure vessel, reactor core, refuelling route and fuel storage facilities, all of which contain regions of high activity.

4.3 The typical difficulties of refurbishing can be illustrated by examining the processes which would be necessary to remove and renew the reactor system of Wylfa nuclear power station. Two main alternatives exist, namely, remove the circulator motors and carry out all dismantling and reconstruction of the system through four circular access penetrations, or re-open a boiler construction access in the top cap of the vessel and undertake part or all of the work through that route. Work could not begin until after the reactor had been shut down for a number of decades and the radiation dose rate had reduced. Corresponding problems can be seen in the multicavity vessels of Hartlepool and Heysham I and in the more recent single cavity vessels of Heysham II and Torness.

4.4 It seems probable that by the time any refurbishment could be carried out the reactor system would be so out of date as to be uneconomic. It is likely therefore that any refurbishing process would also involve the design of a new up-to-date reactor to be fitted within the original void configuration of the old system. Bearing in mind the current proposal for light water cooled reactors in the U.K., the prospect of continuing a generation of gas cooled reactors might be considered wholly unattractive. With such a scenario, the question of refurbishment may only be relevant in practical terms to light water cooled reactor systems.

4.5 Within the limitations of this study it has not been possible to consider in any detail the implications of refurbishing a light water cooled reactor system. However, apart from the reactor pressure vessel itself, the majority of the nuclear steam supply system can be removed and replaced without major demolition of civil structures. This is facilitated by designing some of the internal shielding concrete as demountable panels and by the inclusion of an appropriately sized equipment hatch in the containment wall. In some existing plants, steam generators have already been replaced after plant operating time.

5. PROBLEMS OF DECOMMISSIONING GAS COOLED REACTOR SYSTEMS

5.1 Introduction

5.1.1 Gas cooled nuclear generating plants in commercial operation in the U.K. at the present time are based either on the magnox reactor system or the advanced gas cooled reactor system. The magnox reactor systems utilise single cavity pressure vessels, the earlier ones being of steel construction, the later ones of prestressed concrete. The advanced gas cooled reactor systems utilise either single cavity or multicavity pressure vessels of prestressed concrete construction.

5.1.2 In terms of basic structures, magnox stations and advanced gas cooled stations are essentially similar although the layout of the main buildings varies from station to station. A typical station of either type generally has two reactor vessels housed within a reactor building, the other main structures and facilities common to both types being control room, turbine hall, cooling water pump house and culverts, cooling pond, fuel handling facility and administration block. The respective types of station have other auxiliary buildings to suit their particular functions and requirements.

5.1.3 Site plans of and sections through typical magnox and advanced gas cooled stations are shown in figures 1,2,3 and 4.

5.1.4 The problems of decommissioning gas cooled reactor systems may be conveniently considered in two broad categories. Firstly, there are the straightforward technical problems associated with dismantling and removing a large, complex structure. The fact that some of the structure is radioactive introduces new problems which have not previously been met by many demolition contractors. Secondly, and perhaps more difficult still, are the philosophical problems associated with establishing a common approach to decommissioning in which the attitudes of all parties involved in the planning, licensing, design, construction, operation and demolition can be focussed on the importance and long term benefits of including decommissioning as one of the key parameters from the outset.

5.1.5 These different types of problem are discussed more fully in Sections 5.2 and 5.3 below.

5.2 Technical Problems

5.2.1 In present day magnox and advanced gas cooled reactors the major components of civil works which will cause decommissioning problems are the biological shield, where present, and the prestressed concrete reactor pressure vessel. Such vessels are massive, with walls and caps generally over 5m thick. The two most recent designs of gas cooled reactors in the U.K. are the multicavity style now in the final stages of commissioning at Hartlepool and Heysham I and the single cavity style now under construction at Heysham II and Torness. The former style of vessel is prestressed by a vertical ducted system of tendons in combination with an external hoop wire winding system and contains approximately 30,000 tonnes of concrete. See figure 5. The latter style uses a helical arrangement of ducted tendons and contains approximately 60,000 tonnes of concrete. See figure 6.

5.2.2 During the operating life of a reactor the inner part of the vessel concrete, that is to say, the part close to the reactor cavity becomes activated. This is due to the production of radioactive isotopes of elements in the steel and concrete when subject to neutron irradiation from the reactor core. The extent to which the vessel concrete becomes active depends on the reactor style in question and on the amount of internal shielding placed around the core within the reactor cavity. However, for the purposes of the study, it has been assumed that the activated region behind the liner of the barrel and bottom cap may be up to 1m thick with the top cap being slightly less affected. See figure 7.

5.2.3 It is in the breaking up and removal of this inner layer of active concrete and its embedments and attachments, whilst maintaining the integrity of the structure to retain dust, that the major civil engineering problems of decommissioning the pressure vessel lie.

5.2.4 In considering radiological problems associated with decommissioning it is important to recognise the differences between the activity of a material, the radiation dose a person would receive from being in the proximity of such a material and contamination.

(a) Activity, as stated above, arises from the production of radioactive isotopes in materials subject to neutron irradiation.

(b) Radiation dose received from active material depends on the activity level of the material, the nature of the radiation being emitted, distance from the material, and the degree of intervening shielding.

(c) Radioactive contamination is the deposition of unwanted radioactive material on surfaces and may be partially or totally removed by washing or other decontamination processes.

5.2.5 It is the activity of material that determines the requirements for acceptable means of its final disposal whereas it is the dose rate that mainly determines acceptable ways of removing material from the reactor system.

5.2.6 Within the inner layer of concrete behind the liner of a typical reactor vessel, whilst the overall activity of the concrete will not decay very much during the first 100 years after reactor shut down, the dose rate will reduce substantially during that period. For example, Woollam and Pugh[7] show that to receive a dose of less than 0.5 rem/year, permissible access periods to the inner unlined, unreinforced concrete surfaces of a biological shield for a typical magnox reactor with a steel pressure vessel increase from less than 0.1 hours/week at 10 years to more than 40 hours/week at 100 years after shutdown. The reduction in dose rate from the concrete is mainly due to the decay of short lived gamma emitting isotopes which do not contribute much to total activity but do emit damaging radiation over a relatively short period after shut down.

5.2.7 However, because of the concentration in currently used reactor grade steels of certain trace elements, particularly cobalt, producing gamma-emitting isotopes, the presence of steel in or attached to this inner layer of vessel concrete is a major contributor to dose rates from this region and therefore poses a significant decommissioning problem. Work has been carried out under Project 7 by Boothby and Williams[5] to investigate the control of cobalt content in reactor grade steels.

5.2.8 The presence of active steel in the concrete raises the initial dose rate which then reduces progressively over several decades after shutdown. For a typical magnox reactor, figure 8 illustrates that it takes approximately 100 years for the initially high cobalt (^{60}Co) gamma dose rate to reduce to a level similar to that of the other principal isotopes, namely silver (^{108m}Ag) and niobium (^{94}Nb), and that even at the reduced level, man access to the top and side of the core is limited to approximately 3 hours/week. From 100 years after shutdown onwards, it is anticipated that there will be no significant reduction in dose rate nor increase in access time and this is the reason why current opinion favours waiting 100 years after shutdown before commencing Stage 3 decommissioning but sees no advantage in waiting longer than 100 years. The presence of steel in the form of reinforcement, prestressing tendons and ducts, the reactor liner and its attachments, and penetrations represents a problem to which the study has therefore addressed itself with the objective of being able to reduce this waiting time.

5.2.9 It has been assumed that prior to demolition of any part of the reactor vessel, the reactor internals would have been cut up and removed, most likely through the larger permanent penetrations such as those for gas circulators and man access. This would allow the cutting and breaking up of activated material to take place within an enclosed and controlled environment. An alternative procedure which considers early removal of the standpipe region is discussed in Section 8.5.

5.2.10 The specific identifiable technical difficulties to be faced during decommissioning of existing reactor vessels, and hopefully to be designed out of future vessels, are therefore the conventional problems of demolishing and dismantling large complex structures compounded by the fact that some parts of the structure may be highly radioactive. The main problems are discussed in paragraphs 5.2.11 to 5.2.28 below.

5.2.11 Specialised plant capable of remote working techniques will almost certainly be required for certain aspects of demolition within the reactor cavity. Access provision for this could be considered at the design stage of future vessels.

5.2.12 The selective removal of radioactive material from a massive, partly reinforced PCRV, to leave a structure to which unrestricted access can be allowed for demolition purposes, poses many difficulties.

5.2.13 With the presence of radioactive material, there will be a particular need to control smoke, fumes and dust and to prevent this from escaping to the environment. Various metal cutting techniques such as plasma arc cutting and thermal cutting produce substantial air contamination. Similarly, concrete cutting techniques such as thermic lance or conventional explosive methods produce large quantities of smoke and slag.

5.2.14 As the concrete is progressively broken up there will be the problems of packaging it up in a manner suitable for its removal and final disposal.

5.2.15 Whilst perhaps more of a conventional problem, some care will be needed when removing tendons from within the inner one metre thickness of vessel concrete, as they may have become activated. Removal of wire wound tendons presents further new problems to the demolition contractor.

5.2.16 Since it is possible that the final stages of decommissioning may be delayed until 100 years or so after reactor shutdown, appropriate procedures will be required in connection with long term surveillance of reactor vessels. This not only means keeping a check on the structural soundness of them but also keeping them and the records that relate to them secure over several generations. At the design stage account will need to be taken of the possible extreme conditions of flood, earthquake, aerial impact probability and so on that could occur over a 150 to 200 year period and considerations such as these may provide sufficient incentive for thought to be given to earlier demolition, even if this calls for extensive use of remote handling techniques.

5.2.17 In all existing prestressed concrete reactor vessels, a particular feature that will cause significant problems in decommissioning and dismantling is the requirement for a liner to the reactor cavity. This is required to provide a gas tight membrane around the reactor core and its associated plant and so prevent leakage of material of any kind through the concrete to the environment.

5.2.18 Mild steel is universally accepted as the best material for such a containment liner. It is a material whose properties and characteristics are completely known under all conditions of loading. It is a very durable material of high strength and yet permits easy fabrication into numerous shapes and profiles.

5.2.19 Liner designs are based on all welded fabrications intended to be maintenance free for a working life of 30 years. A typical liner for a PCRV consists of a closed cylinder with flat ends and in existing PCRVs the liner varies in size from 13m diameter and 18m high to 20m diameter and 22m high. Liner thickness also varies depending on the intensity of loading but is generally between 12mm and 20mm. The liner floor is basically a flat plate carrying service penetrations to and from the bottom of the reactor core. In previous designs the steel plate is supported on packs clear of the main concrete foundations and the interspace is grouted up after completion of mechanical installation. Stud anchors are welded to the underside of the plate to effect bond with the grout. The liner roof is also a flat plate, the central position being an integral part of the standpipe assembly.

5.2.20 By virtue of its basic size and the relative thinness of its component make-up a liner requires continuous structural support in all directions in order to withstand the loading from thermal and pressure conditions in a reactor and still remain impermeable. Present liners are therefore usually firmly attached to and fully supported by the concrete structure of the PCRV. Due to this positive attachment, the liner is subjected to the stresses and strains imparted to it by the PCRV during initial prestressing and, later, by movements of the PCRV due to the operating condition of the reactor. To ensure even distribution of these movements across the full size of the liner, attachment of the liner to the concrete is achieved by the embedment of a combination of steel sections and a multitude of studs or hook anchors. See figure 9.

5.2.21 In following the movements of the vessel during initial prestressing the stress in the liner approaches and, in some cases, reaches yield stress. During operation, when temperature has been applied, and in late life, when concrete creep has taken place, the stress in the liner is certainly yield stress.

5.2.22 Due to variation in yield strength of the plate material used to fabricate the liner and also due to normal thickness tolerances, extra support of the liner is needed at the welded junctions. This is achieved by welding onto the back of the liner continuous runs of either angle or channel rolled sections. See figure 9.

5.2.23 All the ancillary services associated with running the reactor core have to be taken into the concrete pressure vessel. This is normally achieved by casting into the concrete, structural steel tubes that are welded to and form an integral part of the main vessel liner system. These tubes, or penetrations, vary in size from a few millimetres up to approximately two metres in diameter.

5.2.24 In some existing reactors, the physical size of the vessel had to be kept to an absolute minimum to satisfy space requirements and this gave rise to problems of shear between the main liner and some of the larger penetrations. This problem was dealt with by incorporating massive steel shear connectors welded to the liner around these penetrations. Discussions with liner designers have suggested that with present knowledge and given the freedom to approach liner design without such requirements, a simplified liner design could certainly be developed. In these discussions, the possibility of using an alternative material to steel was raised. The response was that this had been considered in the past but it had always been found that the functional requirements of the liner are such that steel is the only practicable material to use. Moreover, since the amount of steel needed for a liner is relatively small, it had always been impossible to obtain steel other than that which is commercially available.

5.2.25 Since the concrete and liner are in intimate contact it is necessary to have a cooling system to maintain the concrete at an acceptable temperature. This is achieved by a system of cooling pipes welded to the concrete side of the liner combined with an insulation system on the gas side. The insulation system has to withstand the extreme dynamic loading from the turbulent gas flow and the acoustic response from the gas circulators. It is also subjected to extreme changes of temperature between operating and shutdown conditions. The insulation systems in the most recent gas cooled reactor vessels consist of alternate laminations of an insulating material and stainless steel foil preloaded onto studs, welded to the inside face of the liner. The whole assembly is held in place by individual stainless steel cover plates which are locked on and secured by welding to the studs.

5.2.26 Therefore, the basic liner system capable of meeting the requirements described in the preceding paragraphs is one which is securely fixed to and very difficult to remove from the concrete.

5.2.27 Considering demolition and removal of the liner by conventional means, it would not be practical to tear the liner away from its anchors, bearing in mind that this operation would have to be done from platforms and staging up to 20m high. The liner could be flame cut local to each penetration and anchor, but this method would require previous knowledge of the exact location of each anchor and would require a great deal of thermal energy. The most probable method of demolition would be by blasting the concrete backing to the liner with explosives. If the problem of activity of the liner and the concrete backing behind is added to the general task of demolition of the liner then each individual operation would have to be carried out from the inside until all the active material had been removed. If the level of activity was too high to allow reasonable periods of man access then the whole operation would have to be done by a system of robots under remote control. It is estimated that under these latter conditions removal of a typical PCRV liner would take several years to be completed.

5.2.28 At the end of the decommissioning programme, there will also be the problem of disposing safely of the tools and equipment used for decommissioning as these will have become contaminated during use. This is a problem that is met and dealt with in existing nuclear installations.

5.3 Philosophical Problems

5.3.1 Much current thinking on nuclear power plant decommissioning puts forward the view that adequate techniques are already available for the removal of components and structures from nuclear power plants. Furthermore, with the likelihood that technology will continue to advance, nuclear power plant decommissioning should become less difficult. Against this background it might be difficult for protagonists of "design for decommissioning" to justify extra cost, as would seem to be necessary, of inclusion of design features to aid subsequent decommissioning. Arguments against design for decommissioning would be further strengthened by the pressures that exist to complete the plant and get it operational in the shortest possible time and at minimum cost.

5.3.2 One of the long established design requirements of a prestressed concrete reactor pressure vessel is that it should be designed and built to last, with minimum maintenance, and should resist all known accidents, events and so on that could in anyway damage it and make it unsafe. Such a structure will therefore be very difficult to demolish and remove, particularly when the problem of radioactivity is introduced. In many respects, the introduction of factors to aid decommissioning and dismantling of the civil structure of a reactor vessel may well be in direct conflict with its fundamental design requirements.

5.3.3 It is therefore clear that the introduction of design features to aid decommissioning would require not only the solution of many technical problems but also the conscious making of policy decisions by those involved in the nuclear industry.

6. POSSIBLE FEATURES TO AID DECOMMISSIONING OF GAS COOLED REACTOR SYSTEMS

6.1 Introduction

6.1.1 In this section of the report, every single idea considered as a possible aid to decommissioning is commented upon. Among these ideas are some which, after limited investigation, became less attractive or impracticable, and on which no further work was carried out. On some others which seemed more promising, a considerable amount of work was done and, although still commmented upon in this section, fuller descriptions of such proposals are given in subsequent sections of the report.

6.1.2 The ideas and features that have emerged during the study as possible aids to decommissioning and dismantling of prestressed concrete reactor vessels fall into four main categories. These are:

(a) Planes of weakness.

(b) Features to facilitate forceful break-up.

(c) Selective use and location of materials.

(d) Removal of liner and penetrations.

Within each of these categories a number of separate ideas were developed and these are described in Sections 6.3 to 6.6.

6.1.3 In addition, however, the study has identified one other significant feature that would greatly simplify decommissioning and dismantling of the civil structure of a reactor vessel and that is the introduction of a more comprehensive and efficient shield around the reactor core within the reactor cavity. At one point in the study it was intended to pursue this idea in detail but, on reflection and in discussion with reactor engineers, it was decided that such work fell outside the scope of civil engineering and was not taken any further. However, the background and logic associated with this idea is described in Section 6.2 below.

6.2 Reactor Core Shielding

6.2.1 The literature review and other research has shown that one of the major difficulties in demolishing a prestressed concrete reactor pressure vessel, even 100 years after shutdown, arises from the radiation emitted from the steel components that have become activated during reactor operation. The most effective way of overcoming these difficulties would seem, therefore, to lie in one or more of the following ideas:

(a) Reduce the content in the steel of those trace elements, such as cobalt, which produce gamma radiation. This work has already been mentioned.

(b) Reduce the amount of steel in the region of concrete that becomes active. This will be discussed under selective use and location of materials.

(c) Reduce or eliminate the active zone altogether.

6.2.2 Whilst the work of Boothby and Williams[5] has shown that meaningful reductions in the trace element content can be made, albeit at a price, it is unlikely that complete elimination can be achieved. Additionally, even assuming the elimination of reinforcement and prestressing tendons from the active zone, it would appear from discussions with liner designers that they see no alternative to a steel liner securely fixed to the vessel concrete. There will also always be steel penetrations connected to the liner. Thus the emission of gamma radiation from steel in activated concrete will be a persistant problem to some degree.

6.2.3 It would therefore seem that the most effective way of eliminating the radiological problems is to surround the reactor core completely with sufficient neutron shielding to ensure that the material outside this shielding does not become active. This would have a number of important immediate effects including:

(a) The vessel size may need to be increased to accommodate the extra shielding, although the increase may not be too significant since all-round core shielding is not a great deal more than already provided in existing single cavity vessels. See figure 10. There may be corresponding increases in the reactor building size but these too should be minimal.

(b) There would be more active material to remove from the reactor cavity. However, it should be possible to design the shielding to be in easily handleable sizes and shapes.

(c) The removal of the vessel civil structure would become a demolition exercise free of access restrictions. With regard to the problem of dust, Woolam and Pugh[7] conclude that, even for concrete that had been activated, after 100 years decay, no breathing protection would be required for the concrete demolition work other than that required to comply with the normal safety standards applicable to the inhalation of conventional dust. Some precautions might be needed if activated steel were present in the concrete. However, if the ideal objective of absolute shielding could be achieved, dust problems associated with demolition of the PCRV would be subject to conventional rather than radiological controls.

(d) At an early stage, man access to the annulus between the core shielding and the liner should be possible and this would permit work to commence on removal of plant within this region, liner insulation and the innermost sections of penetrations, and on preparatory work for removal for the liner.

6.2.4 Such an internal shielding structure represents only a little more development of the core shielding arrangement for the Heysham II and Torness reactor vessels for which the activated zone has been calculated to be about 50mm thick. Its design and construction therefore appear feasible and work should be done by reactor engineers to confirm this and to develop the idea for application to future reactor systems. An important aspect of this work will be to assess what effects the incorporation of such a shielding structure may have on reactor operational characteristics.

6.3 Planes of Weakness

6.3.1 An investigation has been carried out into the feasibility of incorporating planes of weakness into the assumed activated zone of a PCRV and between this zone and the non activated material. The object of this is to enable the activated concrete to be broken up into blocks of a suitable size and removed with the minimum of exposure of decommissioning workers to radiation. In achieving this objective however, the structural performance of the PCRV must not be detrimentally altered. A plane of weakness is defined as a plane across which direct compressive stresses normal to it can be transferred but not tensile or shear stresses.

6.3.2 This idea is more applicable to the multicavity style of vessel for which there is no need for man access to the main cavity for boiler service or repair and therefore no need for the same degree of core shielding provided for single cavity vessels. For multicavity vessels, the concrete ligament between the reactor cavity and the boiler cavity serves as boiler shielding and the boilers are removable for service or repair.

6.3.3 The planes of weakness divide the activated zone into blocks and a 1m thick zone has been assumed. It has been decided that these blocks should be of a size such that they can be removed from the PCRV through existing penetrations. This will enable the activated material to be removed using carefully controlled working procedures before any general demolition work on substantially non-activated material is begun. From a consideration of the penetration sizes of the Hartlepool and Heysham I vessel design, and requirements for packaging and transport, it was decided that the blocks should fit into an envelope of 2.0m x 1.0m x 0.6m. Typical barrel and end cap blocks are shown in figure 11. The joints between the blocks - the planes of weakness - are staggered to give the required thickness of concrete for neutron attenuation with no "hot spots" of localised activity formed in the surrounding concrete.

6.3.4 A substantial amount of analytical work has been carried out in connection with planes of weakness and this is described fully in Secion 7. A number of ideas for forming planes of weakness are summarised below and the more promising ones are developed and described further in Section 7.

(a) Place rows of tubes in the concrete on selected planes, possibly making use of suitably arranged cooling water system pipes. See figure 12.

Cooling water pipes are generally provided only on the liner and there is no overall network within the concrete structure. Horizontal holes also cause local loss of shielding and this idea has not been considered further.

(b) Place sheets of plastic, or other material, in the concrete on selected planes, possibly at construction joints.

This idea is discussed in Sections 7.6.2 and 7.6.3. A particular problem, however, with this idea is that some prospective materials are 200 to 500 microns thick and this would be thick enough to inhibit the transmission of prestress to the activated zone.

(c) Inhibit bond at construction joints by painting joints with bond inhibiting material.

The use of materials such as bituminous paint, for example, would not be suitable as the paint layer thickness would be sufficient to inhibit the transfer of prestress. The use of chemical release agents is discussed in Section 7.6.3.

(d) Form crack inducer slots to aid crack formation at controlled spacings when a tensile stress field is induced by use of hydraulic bursters or explosives. See figure 13.

Whilst small crack inducers are effective in forming cracks in relatively thin pavements it is considered that they would not be effective in such a massive structure as a PCRV.

(e) Construct assumed activated zone from precast concrete blocks.

This is discussed in Section 7.6.2.

6.4 Features to Facilitate Forceful Break-up

The ideas comprising this group of suggestions imply building in to the assumed activated zone features which will assist the use of explosive, mechanical devices or chemical agents to cause disintegration of the concrete. These ideas, together with comments on them are summarised below.

(a) Provide preformed holes for the placement of explosive charges. See figure 14.

The use of explosives for the controlled demolition of concrete biological shields and pressure vessels is being investigated by TWC under Project 3 of the CEC programme on nuclear plant decommissioning [2,3,4]. This work indicates that the use of explosives offers an effective method for demolishing massive concrete structures. Additional information on this work is given in Appendix A.

(b) Provide preformed holes for the placement of a chemical demolition agent.

One such agent, suitable for unreinforced concrete, is "BRISTAR" produced by the Onada Cement Company Ltd., of Tokyo, Japan. This product has been successfully used for breaking up rock and massive unreinforced concrete structures. Additional information on its use is given in Appendix A.

(c) Provide preformed cavities for the insertion of mechanical bursters.

Hydraulic bursters can be used effectively for breaking up concrete foundations and it is a reasonable assumption that they would also be effective for breaking up concrete on the inner face of a PCRV provided that a free face had first been established. Further information is given in Appendix A.

(d) Use cooling water system pipes to allow freezing of water and subsequent melting so that expansion will rupture the pipes and concrete. Alternatively pump other fluids along the pipes whose temperature can be cycled from cryogenic to very hot levels.

The same comment as in 6.3.4 (a) applies and, in view of the large amounts of energy that would be required, this idea has not been considered further.

(e) Fill cooling water system pipes with pyrotechnic explosive fluid and use to segment the liner.

This may offer a possible way of breaking up the liner making use of existing pipework and would not therefore require any special additional design features.

(f) Raise concrete temperature to a very high level to reduce strength to aid its break-up.

McFarland[8] has studied the possibility of using heat at temperatures of up to 900°C to help demolish PWR containments. However, application of this technique to PCRVs would involve very large amounts of energy and, for this reason, it is not considered feasible except, perhaps, for localised small areas.

(g) Utilise the high forces of the prestressing system to crack the inner concrete region.

For a wire wound PCRV, it has been calculated that if all the prestress were removed except for the wire windings a little either side of the equator, vertical tensile stresses large enough to cause cracking would develop around the equator level. This technique therefore would have very limited application and is not thought to be feasible for a helically prestressed vessel.

6.5 Selective Use and Location of Materials

The ideas within this group of suggestions are based on keeping the activated concrete as free as possible of materials which are difficult to remove, either because of their tendency to become and remain radioactive for long periods or because of their physical properties and degree of bond with the concrete. The methods that have been considered are summarised below.

(a) Do not place reinforcement or prestressing tendons in the assumed activated zone.

For reasons that have been already stated, the presence of steel, including reinforcement and prestressing tendons, within the activated zone is a severe disadvantage from the point of view of decommissioning.

The Heysham II/Torness PCRV design does not contain significant reinforcement close to the reactor cavity. The prestressing tendons, in solid wall steel ducts, are a minimum of 0.5m away from the main liner and are therefore outside the likely activated zone. See figure 6.

The prestressing tendons of the Hartlepool/Heysham I PCRV design are generally a minimum of 0.915m from the reactor cavity except at the haunches where they deviate and become a little closer. See figure 5. These tendons and their ducts will not therefore become highly activated. This design does, however, for ultimate load purposes mainly, incorporate bonded reinforcement within the activated zone. The effects on the ultimate load factor of either moving this reinforcement to a position just outside of the assumed activated zone or eliminating it altogether are discussed in Section 7.4.4.

(b) Keep reinforcement and prestressing tendons in discrete regions of the activated zone to localise problems due to activated steel.

PCRVs are axisymmetric structures with an essentially axisymmetric stress distribution and reinforcement and prestressing tendons must accordingly be regularly spaced out.

(c) Use for reinforcement an alternative material which does not become highly activated.

Long fibres or mats of fibres of glass, carbon or aramid have been used to reinforce cement paste and concrete but are not considered practical for use in the large concrete pours associated with a PCRV. Short chopped strands could be mixed with the concrete but these would not fulfill the main function of reinforcement close to the reactor cavity, namely to control the opening of cracks, particularly at ultimate load.

(d) Develop a steel with minimum content of trace elements which form long half-life radioactive isotopes.

This topic is being studied by others, notably Boothby and Williams[5].

(e) Use a type of concrete which can be made to degrade when subjected to chemical or physical process not experienced during reactor operation.

A property of normal concrete is that it is an inherently durable material. Some chemicals react with the cement paste and break it down but such massive quantities of these agents would be required that such a process would not be practical for a PCRV. The degradation of concrete by heat, freezing and temperature cycling has been discussed earlier but these processes would be impractical for the large scale breaking up of a massive concrete structure.

(f) Provide a non-structural layer of a removable material on the inside of the liner to shield the structural concrete.

This is similar to the idea discussed in Section 6.2.3.

(g) Use preformed steel blocks or blocks of some other material to replace concrete as a structural and shielding material within the activated zone.

The undesirability of steel in the activated zone has already been discussed and problems associated with the use of preformed blocks are described in Section 7.6.2. Furthermore, no material has been identified which combines the properties of high stiffness and compressive strength, economy, ease of placing and good radiation shielding exhibited by concrete.

(h) Improve the neutron shielding properties of the structural concrete adjacent to the reactor cavity to reduce the volume of activated concrete, and reduce the surface dose rates.

Increasing the amount of hydrogen in concrete improves its neutron attenuating properties, as demonstrated by Dyson and Harrison[9]. Two possible methods of increasing the proportion of hydrogen in hardened concrete are to increase the amount of free and chemically bound water in the hardened cement paste or to use hydrous aggregates.

The presence of boron in concrete has been recognised as a means of improving the effectiveness of concrete as a neutron shield and reducing its activity level[10]. By using non-soluble boron frits, researchers in Japan[11] have succeeded in producing concrete mixes with a boron content of up to 6.4%.

Further discussion of the above two ideas is contained in Appendix A.

6.6 Removal of Liner and Penetrations

This group of suggestions relate to features or arrangments which will minimise, or eliminate, in future PCRVs, the problems that will have to be overcome in removing the liner and its associated penetrations from existing PCRVs. The ideas fall into two main categories, namely, those that may be considered for existing liner designs and those that envisage major changes in concept and approach to liner design. The suggestions within the former category are described below while those in the latter are described in full in Section 8. Removal of the standpipe penetrations is a particular problem also described in Section 8. The ideas that have been considered are discussed below.

(a) Reduce the level of activity of the liner by providing a non-structural shield wall between it and the core.

This is essentially the core shielding idea described in Section 6.2.3 of the report. It is unlikely that such a shield would guarantee a level of no activity in the liner. Local weaknesses, particularly at penetration locations, would have to be dealt with individually but, in general terms, it should be possible to provide for direct man access to the liner. The introduction of shielding would not, of course, alter the total inventory of activated material. It would simply transfer the problem of breaking most of it up and handling it from a monolithic structure to a fabricated one which should be easier to deal with.

(b) Use an alternative material to mild steel for the liner.

It has already been stated that mild steel is the best material for the formation of an impermeable containment liner and all enquiries to date have failed to identify a better material. For further information, see Appendix A.

(c) Alter stud or hook anchor details to enable individual removal of them from the liner. See figures 15 and 16.

A loose bolt type of fitting for the liner anchors would not be acceptable to liner designers because it would depend on a nut and washer to seal the preformed hole in the liner. Liner anchors penetrating and projecting from the liner face would not be acceptable either since the resultant thickness of liner around the perimeter of the preformed hole is only the thickness of the fillet weld to the anchor rod. The detail could be changed to give, say, a full penetration weld but the cost would then increase significantly.

(d) Alter the detail at liner to penetration junctions to enable independent detachment and removal of each. See figures 17 and 18.

Removal of the liner and penetrations would be eased if a short length of penetration immediately behind the liner were removed first. Referring to figure 17, this could be achieved by encasing, say, one metre of the penetration in concrete with stepped joints for shielding and a slight taper for withdrawal of this plug from the body of the PCRV. The surface of the works-placed concrete collar would be coated with a debonding agent. Removal of the plug would be achieved by internally cutting through the penetration liner at the bottom of the collar, cutting the main liner around the circumference of the collar and pushing the freed mass into the vessel by jacking from the outside.

Grouping penetrations together in clusters, as shown in figure 18, and so concentrating decommissioning problems into precise areas, is not considered feasible since the nature of a reactor core generally requires that pipes and services radiate out in a balanced configuration.

(e) Inject explosive material into the liner cooling system pipework and blast liner away from concrete.

This idea has been commented upon in Section 6.4 (e).

(f) Avoid the need for massive steel fabricated anchors.

This is a matter of design detailing dependent on layout and space requirements.

(g) Incorporate lifting beams around the perimeter of the liner roof soffit.

These would provide suspension points for demolition equipment and could also be used during construction to assist, for example, in boiler installation.

7. PLANES OF WEAKNESS

7.1 Introduction

7.1.1 In Section 6.3 of this report the idea of incorporating planes of weakness within the assumed activated zone of the PCRV was briefly described. In the paragraphs that follow, the work that has been done to assess the feasibility of this idea, both with respect to design and construction aspects, is described in full.

7.1.2 The structural implications of introducing planes of weakness into the activated zone of a typical PCRV have been assessed under service load conditions, seismic loading and ultimate load.

7.1.3 Studies were based on the Hartlepool reactor vessel, firstly because it is to this style of vessel that the concept would primarily apply and, secondly, so that reference could be made to the known behaviour of this vessel without planes of weakness. Two structural configurations were considered, the first, structure type 1, being shown in figure 19, and the second, structure type 2, in figure 20. Structure type 1 is the same as the original vessel, with no planes of weakness. Structure type 2 is the original vessel but with planes of weakness in the bottom cap and barrel. No planes of weakness were introduced into the top cap because of:

(a) the difficulties of dealing with gravity loads.

(b) difficulties due to the large number of standpipe penetrations.

7.1.4 The layout of the planes of weakness was chosen to take into account the need for:

(a) Structural integrity.

(b) Effective neutron shielding of the outer concrete.

(c) Ease of removal of blocks from the reactor cavity and their ultimate disposal.

7.2 Service Load Conditions

7.2.1 Structures type 1 and 2 were analysed under service loading by means of a finite element axisymmetric analysis. The grid for the analytical model is shown in figure 21. The structures were subjected to the following load cases:

(a) Load Case 1 - initial prestress.

(b) Load Case 2 - prestress plus proof pressure.

7.2.2 For the purpose of interpreting and comparing results, a number of control points in the structure were selected and these are shown in figure 22. The results for structure type 1 were verified by comparing them with results from an analysis of the Hartlepool PCRV carried out at the contract design stage. See figures 23,24 and 25. Results of the analysis of structures type 1 and 2 were compared on the basis of stresses at the selected control points as shown in figure 22. Stress profiles have been plotted along all the lines shown for both load cases and twelve representative comparisons are shown in figures 26 to 37.

7.2.3 It will be observed that the overall stress pattern in the two structures is substantially the same. Examination of the full set of forty-two comparisons indicates some small differences in the stress plots for profile 7 under load case 1. Under load case 2, however, these differences are almost completely cancelled out. Using the criteria in BS 4975[12], the stresses for structure 2 would be within acceptable limits.

7.2.4 Although not evident from the stress profiles, tensile stresses occur around the haunches of both structures under load case 2. These were looked at closely and were found to be small and localised. At the lower haunch of structure 2 they are localised within the inside corner block. The tensile zones that occur fall well within the limits laid down in BS 4975[12].

7.2.5 This analysis showed that the behaviour of the vessel under service loads is very similar with and without the chosen arrangement of planes of weakness.

7.2.6 The analysis did not consider the effects of temperature loading. The general effect of operational temperature conditions is to add compressive vertical and hoop stresses to the inside of the barrel and tensile vertical and hoop stresses to the outside. It is considered that the inclusion of planes of weakness would not significantly affect the stress distribution due to temperature loading. It should be noted that small tensile stresses occur close to the liner during PCRV cool down procedures. However, it is expected that these would be accommodated within the thin layer of reinforced in-situ concrete immediately behind the liner referred to in Section 7.5.

7.3 Seismic Loading

The stability of discrete blocks formed in the wall of the vessel has been examined. Simple calculations have shown that the blocks have a high degree of stability under seismic loading based on criteria used for the design of the Heysham II and Torness power stations. The presence of a thin layer of reinforced concrete immediately behind the liner would further enhance this stability.

7.4 Ultimate Load Conditions

7.4.1 The ultimate load characteristics of structures type 1 and 2 were obtained using an in-house computer program. This program examines a number of possible failure mechanisms in order to find the one associated with minimum potential energy of the vessel. The mechanisms are deflected in increments, and at each step the vessel pressure which would maintain the mechanism in equilibrium is found. Shears, deflections, prestress strains and crack sizes are output at each step so that the progressive plastic behaviour of the vessel may be monitored. Three failure criteria are recognised and overall vessel failure is assumed to occur when one of the three is violated.

7.4.2 The three criteria for failure are:

(a) Tendon failure. The vessel is assumed to fail if the strain in one of its tendons exceeds its ultimate strain. This is taken as the strain corresponding to 90% of the guaranteed ultimate tensile strength (GUTS).

(b) Failure of the concrete in either compression or shear.

(c) Excessive crack widths at interfaces with liners. It is assumed that the liners can span a crack width of not more than 32mm at ultimate load conditions and that a larger crack width will cause an unacceptable liner failure, leading to a "gas in cracks" condition.

7.4.3 In modelling structure type 2 for the ultimate load analysis it was assumed that the planes of weakness zone could not carry tension. Any reinforcement which was in this zone in the original structure was moved to a position just outside the planes of weakness zone.

7.4.4 Results of the ultimate load analysis are presented in table 1. It will be seen that for structure type 2, a 20% increase in the area of bonded reinforcement at the haunches (now moved outside the planes of weakness zone) is necesary to achieve the same load factor as structure 1. In each of the three cases listed in table 1 the ultimate load is determined by the longitudinal barrel crack width reaching the allowable limit at the interface with the vessel liner. Further work was carried out to determine the ultimate load behaviour of the vessel with all the bonded reinforcement removed. It was found that to achieve the same factor an increase in the quantity of prestressing steel of over 30% was required. Whilst this might ease the decommissioning problem it would be very expensive. The "gas in cracks" condition has not been considered.

7.5 Attachment of Liners

In conventional PCRV design the integrity of the main liner and other liners, if applicable, demands that they are anchored by means of short steel studs or hooks some 150mm to 300mm long into a continuous backing of in-situ concrete. It would not therefore be feasible to extend the planes of weakness zone right up to the liners. It is proposed therefore that the liners would be backed with a continuous layer of reinforced concrete some 200mm to 350mm thick which contains no planes of weakness but which is not bonded to the planes of weakness zone. This would stiffen the liner against internal buckling but would not represent an excessively thick layer to be stripped off. The planes of weakness zone would also have to be stopped short of penetrations through the structure, leaving them bedded in at least, say, 150mm of continuous in-situ concrete construction.

7.6 Formation of Planes of Weakness

7.6.1 Two basic forms of construction for the planes of weakness zone have been considered, namely, precast concrete construction and in-situ concrete construction. The objective of both forms of construction is that the activated zone shall effectively be constructed from blocks which can ultimately be dismantled and removed from the reactor cavity with the minimum of exposure time to decommissioning workers. However, the form of construction has to be such that the structure performs in essentially the same way as conventional forms of PCRV construction.

7.6.2 In so far as precast concrete construction is concerned two variations of this idea have been considered in detail, namely, close tolerance blocks butted together or blocks bedded in mortar.

(a) Close tolerance blocks butted together.

This would involve blocks like those in figure 11 cast in high quality moulds to very small tolerances and butted togethei with dry joints. It is unlikely, however, that tolerances better than $\pm$ 1mm could be achieved on the dimensions of the blocks. A consideration of strains under initial prestress has shown that a planes of weakness zone constructed of such blocks would not satisfactorily distribute prestress. For example, the hoop strain under initial prestress at the equator at the outer surface of the planes of weakness zone would be approximately 2.7 x 10^{-4} or 0.27mm in 1m. Clearly, because of the relative magnitude of the tolerances, prestress would only be induced at contact areas at the local "high spots" on radial planes and the region could therefore be largely unstressed. If this were the case the planes of weakness zone would merely transfer the gas pressure by means of radial stresses to the outer part of the structure and would not contribute to the overall structural action of the vessel.

Consideration has been given to lining the blocks with steel or some other material. It is not considered, however, that this would improve the accuracy of the blocks sufficiently. Furthermore, the presence of additional steel in the activated zone would firstly be expensive and secondly be undesirable from radiological considerations.

(b) Blocks bedded in mortar.

The problems referred to in (a) above would be overcome if the blocks were separated by joints filled with mortar. The blocks would probably have to be smaller than those used with dry joints in order that during placing the mortar did not bleed excessively from the joints due to the weight of the blocks. The optimum size of block would have to be determined by trials and it may be that blocks small enough to be placed by hand would prove to be the most economic. In order to produce the planes of weakness required to facilitate decommissioning, half of the block faces would need to be treated with a chemical release agent to inhibit bond with the mortar. The mortar would be applied to the remaining untreated block faces, which should be clearly marked. Due to the level of compressive stress under service and ultimate load conditions it might be necessary to use epoxy mortar rather than conventional building mortar. This would only be suitable if a release agent were available which would be effective for epoxy mortar.

7.6.3 In so far as in-situ concrete construction is concerned, again two methods for forming the planes of weakness have been considered. These are, firstly, normal sized pours with special formers placed within the formwork before placing concrete and, secondly, small pours with bond inhibitor at construction joints.

(a) Normal sized pours with special formers.

A variety of sheet materials have been considered as planes of weakness formers, the object being to find a material which has the following properties:

- sufficient strength and stiffness to withstand loads from wet concrete.

- ability to transfer compressive stresses normal to the plane of the sheet.

- lack of bonding to hardened concrete.

- fairly easy to handle and place.

The most promising sheet material identified would be a 10mm to 20mm thick sheet of fibre reinforced cement. This would have the necessary strength and stiffness properties and if coated on one or both sides with a chemical release agent would promote the required planes of weakness. Fibre reinforced cement sheets using alkali resistant glass fibres are in common use for purposes such as cladding. Alternatives, albeit more expensive ones, would be sheets reinforced with fibres of aramid or carbon, or sheets of high density polyethylene.

Such sheets could be used to form vertical planes of weakness within a pour. A plan on a typical pour is shown in figure 38. Horizontal planes of weakness would be formed by preventing bond at horizontal construction joints by leaving a smooth finish to the joint and applying a chemical release agent.

(b) Small pours with bond inhibitors at joints.

This alternative would save the cost of providing and fixing bond inhibiting panels but would involve a larger number of smaller pours. Finished surfaces of the pours would be left smooth and treated with a chemical release agent to prevent bond with the adjoining pour. The economics of a large number of small pours would need to be considered carefully but the use of high quality steel forms would enable rapid fixing and re-use many times. Work could proceed on a number of pours simultaneously.

8. REMOVAL OF LINER AND PENETRATIONS

8.1 Introduction

8.1.1 In Section 6.6 of this report some ideas that would entail minor changes to existing liner design details, to aid decommissioning, were discussed. The paragraphs that follow discuss some possible, more wide ranging changes that could be made in liner design philosophy to facilitate the eventual removal of the liner from the vessel during the decommissioning phase.

8.1.2 Within the scope of the study it has been possible only to discuss such conceptual changes in principle without significant attention either to detail or to possible consequential effects on nuclear engineering aspects of the reactor system. Such factors can only be assessed through a more extensive feasibility study.

8.1.3 The conceptual ideas proposed are discussed in paragraphs 8.2 to 8.5 below.

8.2 Freestanding Liners

8.2.1 The demolition and removal of a liner for a PCRV would be considerably simplified if it was possible to have a liner that was largely detached from the concrete vessel.

8.2.2 Free standing liners have been considered in the past[6]. Major problems with this concept are the liner being able to withstand vessel pressure and to accommodate movements across penetrations attached to both the liner and the concrete vessel. The first of these problems could be overcome by pressurising the space between the liner and the vessel concrete.

8.2.3 A PCRV for a commercial nuclear installation is fundamentally an immensely strong and massive structure. It was therefore natural for design engineers to utilise the strength of such a structure to carry the forces on the liner rather than provide an alternative comprehensive support system.

8.2.4 Perhaps the design philosophy for future installations should be to continue using the PCRV as a means of support for the liner but in some way free the liner to allow independent movement and subsequent easier dismantling on decommissioning.

8.3 Grouted Liners

8.3.1 The design development of steel liners for PCRVs over the years has led to the currently acceptable principle of bonding the liner to the concrete with a multitude of closely spaced stud or hook anchors.

8.3.2 The main requirement for the stud or hook anchors is to prevent the liner from buckling away from the concrete, principally during initial prestressing and subsequent creep of the vessel concrete when the compressive stresses in parts of the liner may approach yield stress. The tendency for such buckling to occur may also be aggravated by temperature effects during reactor operation and the long term creep of the concrete under prestress loading.

8.3.3 It would therefore be advantageous if some means could be devised for isolating the liner from the effects of this prestress loading and as much as possible of the prestress induced concrete creep. If this could be achieved, the compressive loading in the liner due to temperature and residual concrete creep might be low enough to eliminate the buckling problem or reduce it to an extent such that it can be catered for by local stiffening or by only a minimal number of liner anchorages.

8.3.4 One possible method of isolating the liner from the concrete vessel until some stage after vessel prestress is applied is the formation of a small construction gap between the liner and the vessel concrete. This gap would be approximately 300mm wide and on completion of the prestress period would be grouted up. See figure 39.

8.3.5 There are two ways in which this gap could be formed. Either erect the liner first and construct the vessel walls, including gap, or construct the vessel walls first and lower the liner in position maintaining the gap.

8.3.6 The first method follows the current approach to construction of a PCRV. The liner is erected on the vessel bottom cap with the penetrations attached. The vessel walls are constructed using the liner to support formwork for the internal face. Formation of the construction gap could be achieved by using a permanent or built in shutter. The liner would be stiffened by continuous rolled sections and these could be used as support points for the shutter. If this shutter was made out of prestressed planks or precast concrete blocks, this form of construction could also be associated with the "planes of weakness" proposal.

8.3.7 The second method involves conventional construction of the pressure vessel walls using removeable formwork to both faces. Through wall penetrations would be cast in and could be accurately located by making the internal wall shutter a template. When the walls are complete to the underside of the top cap level the liner would be lowered in one piece into the void. Connection between the liner and penetrations would be done after prestressing and could be as shown in figure 40.

8.3.8 Both of the above methods take the construction to top cap level and in this position it is normal for the outer ring of the liner to be included in the first stage of liner construction. The inner circle of liner roof would be suspended with the standpipe assembly and the whole unit concreted in with the remaining walls up to the top prestressing galleries.

8.3.9 For concrete under load it is estimated that of the 30 year creep, 50% will have taken place after three months and 80% after three years of the load being applied.

8.3.10 The prestressing of a typical PCRV would take about nine months to complete and of course some of the early life creep will take place during this period. If the grouting operation was delayed for six months after completion of prestress this would eliminate about 70% of the strain due to long term creep on the liner.

8.3.11 It is acknowledged that the proposal of forming a construction gap between the liner and the vessel poses many new problems from both a design and construction point of view. Both construction methods mentioned above in some way reflect what is currently acceptable to design and site engineers. Present day liners have internal bracing to prevent distortion during the construction period, but this bracing allows the gas baffle, where appropriate, to be lowered into the central core before the liner roof is completed. This form of construction could be maintained and as far as lifting the liner into position is concerned, the all up weight of the liner is similar to the weight of the gas baffle structure.

8.3.12 Allowing for any residual movement between through wall penetrations and the liner would be one of the main problems to solve. These connections could be done at a late stage by means of floating spigot pieces as shown in figure 40. Alternatively, double skin penetrations could be used and the interspace grouted up at the same time as main liner grouting.

8.3.13 Another important problem would be in sealing the connection in the liner roof between the inner circle part of the standpipe assembly and the outer ring of the liner roof. The physical connection is no different from that which is currently made on the liner floor. Due allowances would have to be made on site in the initial setting of the level of the inner ring to accommodate all the movements up to the completion of prestress and initial allowance for creep.

8.3.14 It is envisaged that there will be problems in connecting up the cooling water pipes, particularly in the scheme where the liner is lowered into the vessel. However, for the purposes of this report, this problem is considered one of detail only and should be tackled in the working design stage. The basic principle of using small diameter pipes welded to the back of the liner would still be applicable.

8.3.15 Having accepted the basic principle of a grouted liner, an assessment would have to be made of what, if any, liner anchorage would be required to prevent liner buckling as a result of the residual concrete creep and temperature loading during reactor warm-up and operation. If it is proven that operational strains can be accommodated by conventional stiffening techniques, such as welded on rolled sections, then the use of blanket coverage of hook anchors would no longer be necessary. The liner would always be anchored at points of through wall penetrations and it might be considered prudent to have a single row of intermediate anchors where the vertical height between penetrations is considered excessive.

8.3.16 If additional anchorage of the liner is not required, the function of the grout is reduced to one of providing a means of transferring the direct load on the liner from the gas pressure onto the inside face of the concrete pressure vessel. In this respect the grout acts purely as a filler. In keeping with the philosophy of eliminating steel from the inner metre of vessel concrete this band of grout could be unreinforced. A further advantage may be accrued by debonding this grout from the vessel concrete.

8.3.17 An alternative way of dealing with the effect of heat on the vessel concrete might alleviate some of the problems associated with the currently accepted method of insulation attached to the inside face of the liner. This situation might be achieved if a heat resistant grout could be developed. The possibility of producing such a grout could only be established by laboratory trial mixes and testing. Further thoughts on liner insulation are given in Section 8.4 below.

8.3.18 Another possible development in the vessel cooling system would be to locate a layer of conventional insulating material between the grout layer and the vessel concrete, say by fixing this material to the concrete face of the vessel before liner installation and grouting. This scheme has the advantages of buried, and therefore, protected insulation. The insulation debonds the grout and could be considered in conjunction with the planes of weakness philosophy. The major disadvantage is the effect of high temperature directly on the liner.

8.4 Liner Insulation

8.4.1 The need to protect the inner layer of concrete of a PCRV from the excessive heat generated by the reactor core has led to the development of a highly efficient but very intricate insulating system. The use of stainless steel and the welded fabrication of such a system complicates the decommissioning of a liner in active areas.

8.4.2 Some degree of failure was experienced in early insulating designs primarily due to dynamic loading and vibration from the gas circulators. It is possible that designers of gas circulators could benefit from the experience of the aircraft industry where, in future, quieter jet engines will be required by law. Development of the fan blade design might reduce noise levels and consequent vibrations. This in turn might allow some relaxation in the degree of fixity of the insulation.

8.4.3 Looking again at the insulation complexities from a purely logical point of view, the best way of eliminating this problem would be either to replace the insulation with an alternative method of cooling or to move the insulation away from the internal face of the liner. In the first instance it is a fact that high temperature furnaces are insulated with either fireclay bricks or a moler concrete lining. It is possible that a similar system might be developed for the back of the liner. In the second instance it is possible that the insulation might be used in a sandwich construction with a double skin liner. It should be pointed out that in both of these cases the temperature of the internal face of the liner would be increased. The effect of this on the integrity of the liner in operating conditions can only be assessed by design engineers in a full working design of a PCRV.

8.5 Removal of Standpipes

8.5.1 The standpipe assembly is contained within a 10m diameter circle of the pressure vessel top cap. It is central about the vertical axis of the reactor core and is the means of access for fuelling and defuelling the reactor. Provision is also made in the assembly for service penetrations used in monitoring and moderating the operational condition of the reactor.

8.5.2 Each standpipe is generally made up of a system of concentric tubes, mild steel being used as a liner and stainless steel as the internal guide tube(s). Insulation is contained between inner guide and liner tubes and cooling is effected either by a system of helical cooling pipes or by a water jacket. The standpipes are arranged in a closely spaced symmetrical grid and in a typical assembly there would be four fuelling standpipes with a central control rod standpipe within an area of approximately 1 square metre on plan. The bottom end of each standpipe penetrates through and is welded to the roof of the pressure vessel liner. In this area the thickness of the liner is increased primarily to withstand the construction loading from wet concrete and secondly to be rigid enough to act as a template for accurate location of each standpipe. The whole assembly is encased in unreinforced concrete for a depth of approximately 5.5m.

8.5.3 The extensive fabrication of the standpipe assembly is completely attached to the liner roof and therefore ensures that the latter is securely anchored to the top cap concrete of the pressure vessel. Any attempt at dismantling this area of a PCRV would automatically have to take into consideration some method of detaching or dismantling the standpipe tubes.

8.5.4 The liner and the lower portion of standpipes would almost certainly be active in this region. Whatever amount of shielding is placed on top of the reactor core below the gas baffle dome, there will always be shine paths through the guide tubes between the standpipes and the reactor core. It is also anticipated that the full height of each standpipe would exhibit some degree of activity and contamination at the end of the working life of the reactor, due principally to deposition of spalled oxide from fuel cladding and active particles of dust.

8.5.5 The layout of the standpipe assembly is a function of the working of a nuclear reactor and it is not envisaged that this layout could be changed in any way purely to facilitate decommissioning of a PCRV. The standpipe region has to withstand the internal pressures of a PCRV and consequently its structural depth has to be consistant with the required minimum structural thickness of the PCRV.

8.5.6 Activity levels, particularly in the lower portion of the standpipes, would require some or all of the dismantling in this area to be carried out by remote control methods.

8.5.7 Whilst the proposed method of dismantling a reactor core is not within the scope of this study it would seem that, from a practical demolition point of view, control of the dismantling of the reactor core would be greatly facilitated if a substantial overhead means of access were provided. This could be achieved if the entire standpipe assembly were removed thus providing a 10m diameter access directly over the core.

8.5.8 Dismantling of both the standpipe assembly and the core could be done by remote control using the same mechanical plant and equipment. After shutdown of the reactor and removal of the fuel a temporary shielded cover would be erected over and clear of the standpipe tops at operating floor level. See figure 41. This cover would probably be a precast, prestressed concrete slab and would be supported on the outer ring of the top cap structural concrete with side walls as required. The new chamber now formed would be completely shielded and provided with access for men and plant. The underside of the cover would be fitted out with runway beams, lifting tackle, remote control mechanisms, TV monitors and so on, all of which could be assembled and tested in advance of dismantling any active material. The route out for active material would be by shielded flask loaded under the temporary cover and fed out on a rail through a shielded ante-chamber to the reactor operating floor level.

8.5.9 To accommodate this method of dismantling the standpipe assembly and core, the ring of concrete around the standpipe zone would have to be designed to be self supporting on removal of prestress and the standpipe assembly. This ring of concrete would also have to carry the imposed load from the temporary concrete cover and any plant loading during dismantling down to the bottom of the reactor core.

8.5.10 On a twin reactor station the temporary cover could be re-used by lifting it over to the second reactor on completion of dismantling the first reactor.

8.5.11 The standpipe assembly would be dismantled from the top downwards and the work could be carried out in three stages. See figure 41. The first stage would be complete removal of all steel and concrete down to a depth of 2.5m below fuelling level. The second stage would remove the next 2m of the assembly and the third stage would remove the remaining 1m of the assembly together with the roof of the PCRV liner. The third stage would involve handling the most active materials.

8.5.12 In the first two stages each standpipe would be considered individually. A circumferential horizontal cut would be made through each standpipe at each stage level and a vertical cut would be made from the top to the bottom of the second stage. Cutting would be done by a flame or machine cutter located in the bore of the standpipe and the length of standpipe in each stage would be extracted from the concrete by jacking against the infill concrete. The success of this operation would depend on each standpipe having a reasonably smooth external profile and this would be achieved if the water cooling was contained within a cylindrical jacket and not by helical pipe as currently used in some designs.

8.5.13 The infill concrete would be removed in sections by breaking the concrete across the narrow ligament between standpipe holes by hydraulic or mechanical bursters. Sections of the perforated concrete could be lifted out at a time and this operation would be facilitated by having a horizontal construction joint in the concrete at the first and second stage levels.

8.5.14 For the third and final stage of dismantling it would be proposed to operate on a modular basis, that is to say, remove the five standpipes and the metre square of liner at the same time. Each module of standpipes would be encased in a 1m deep plug of works placed concrete immediately on top of the liner. The gaps between concrete plugs for each module would be grouted up. Dismantling would be carried out by cutting the liner from the underside into one metre squares and cutting the standpipes between the liner and the gas baffle. Each plug of concrete would then be jacked out. This operation would have to be carried out radially outwards from the centre.

8.5.15 Removal of the standpipe region in this way then provides shielded access to the reactor core.

8.5.16 An alternative option for dismantling the standpipe region, to give access to the reactor core, would be to design all or part of the standpipe region as a separate removable component. This would require the provision of equipment capable of lifting a prestressed concrete disc of up to 10m diameter and 5.5m thick, penetrated by several hundred standpipes. Such equipment could however be designed to serve many purposes during both construction and operation of the reactor. Variations on such a scheme have been considered in the past by reactor vessel designers and, although thought likely to be feasible, none of the postulated options has been developed in detail.

9. DECOMMISSIONING OF LIGHT WATER COOLED REACTOR SYSTEMS

9.1 Introduction

9.1.1 This section of the report defines alternative procedures by which the decommissioning of a 1200 MW(e) pressurised water reactor (PWR) could be undertaken. A number of scenarios are postulated and the decommissioning process associated with each scenario is described. Where appropriate, references are given which cover specific aspects of decommissioning in more detail. Sufficient description of both the main plant items and the major civil structures is given to enable meaningful statements on various aspects of decommissioning to be made.

9.1.3 This section of the report also comments upon existing design or construction features of typical PWRs which ease the problems of decommissioning. In addition, new design features which, if incorporated into future PWRs, would simplify decommissioning are identified and discussed.

9.2 Typical PWR Station Layout

9.2.1 Major Structures and their Functions

The layout of a typical PWR station, as shown in figures 42 and 43, is based on the Standardised Nuclear Power Plant System (SNUPPS) design developed in the USA during the 1970s. The structures of primary interest during decommissioning are those which contain radioactive material or fluid systems. The other structures present only demolition problems without radiological implications. Hence the major structures having radiological implications are as follows.

(a) Reactor Building

The reactor building houses the Nuclear Steam Supply System (NSSS), described in Section 9.2.2. This structure takes the form of a right vertical cylinder capped by a hemispherical dome and has an essentially flat foundation, as shown in figure 44. The shell structure is constructed from post-tensioned, prestressed concrete with a reinforced concrete foundation slab and the internal surfaces of the shell and foundation are steel lined. The shell structure has numerous small penetrations for piping and cabling, plus air-lock type man access penetrations and a plant access penetration. The reactor building shell and foundation form a containment barrier which serves as biological shielding and protection for the NSSS during normal operation and would contain any spilt contaminated coolant during a pipe-break fault condition. Major reinforced concrete structures within the reactor building are shown in figure 45.

(b) Radioactive Waste Process and Storage Building

The waste process building is of conventional, thick section, reinforced concrete construction but with some structural steel beams which support in-situ floor slabs. The process plant is housed in a complex arrangement of internal cells, some of which are formed in concrete blockwork. The building has four principal floor levels and a built up roof supported by structural steelwork. The adjacent drum storage and low activity waste treatment building is of similar but lighter construction having three principal floor levels. The radwaste pipe tunnel connecting the auxiliary building to the waste process building is a reinforce concrete two cell box structure.

(c) Fuel Building

The fuel building is a rectangular, reinforced concrete structure containing a reinforced concrete, stainless steel lined fuel storage pond for underwater storage of new and irradiated fuel. Separate stainless steel lined compartments are provided for underwater transfer of new and irradiated fuel and to allow loading of irradiated fuel into the transport flasks. The intermediate floors within the building and the roof are reinforced concrete slabs supported by structural steel beams.

(d) Auxiliary Building

The auxiliary building is essentially of reinforced concrete construction with one side formed by the curved wall of the reactor building. The floor and roof slabs are supported by a structural steel frame. The building houses essential systems to ensure safe reactor operation and these are, mainly, the chemical and volume control system (CVCS) which controls the boron content and chemistry of the primary circuit water, the residual heat removal systems (RHRS) to remove decay heat from the reactor when it is depressurised, and the component cooling water (CCW) system which is used to transfer heat from the RHRS and other reactor auxiliary plant to the sea water system.

(e) Control Building

The control building is of reinforced concrete construction with the floor slabs carried by a structural steel frame. The building contains the main control room, the associated data processing equipment and the reactor protection system central equipment. An emergency control room, which includes an auxiliary shutdown panel, is provided.

(f) Turbine Building

The turbine building houses two parallel turbo-generator sets each with its own auxiliaries. It is of reinforced concrete and structural steel construction.

(g) Other Structures

The remaining structures shown in figures 42 and 43 are of conventional construction.

9.2.2 The Nuclear Steam Supply System (NSSS) - Figure 46

The Nuclear Steam Supply System based on the standard Westinghouse four-loop pressurised water reactor produces 3425 MW of heat. The reactor is cooled by water pressurised to 158 bar (2250 psia) which is pumped through four closed primary loops each of which contains a steam generator and a pump. The steam which is produced in the secondary side of the steam generators is passed to two 600 MW turbine-generators which produce a combined nett output from the power station of 1110 MW(e).

(a) Reactor Pressure Vessel (RPV) - Figure 47

The RPV consists of a vertical steel cylinder closed at each end by a domed head. The lower head is integral with the cylinder, while the upper head is removable and is retained by bolts.

The cylindrical section comprises a lower, plain portion which is slightly thickened in the region of the four inlet and four outlet nozzles, arranged symmetrically around the vessel. The inlet nozzles connect the vessel to the reactor coolant pumps (RCP) and the outlet nozzles connect the vessel to the steam generators (SG) via the reactor coolant piping. Above the nozzle region is a flange penetrated by holes for bolting to the closure head. The internal diameter of the flange is machined to provide a ledge from which the core and vessel internals are supported. A cavity seal, which is required to contain the refuelling pond water during refuelling, is attached to the outside of the flange. All of the internal surfaces of the RPV are in contact with reactor coolant and are therefore clad with corrosion resistant materials, either stainless steel or a nickel alloy with composition equivalent to Inconel alloy 600.

Approximately once a year the reactor is shut down for refuelling. At that time, the reactor vessel head is removed and one third of the fuel assemblies are removed underwater and taken from the containment, still underwater to a water filled storage pond in the adjacent fuel handling building. The fuel remains in this building until its radioactivity and temperature have decayed to a level at which it can be transported off site. The remaining fuel assemblies are re-arranged within the core and new fuel assemblies are inserted.

(b) Reactor Vessel Upper and Lower Internal Structures

The RPV internal structures comprise a lower assembly which supports the core and an upper assembly which locates the top of the core and guides the control rods. During operation both the upper and lower internal assemblies are completely enclosed within the reactor vessel and are supported from the area of the main vessel flange. To refuel the reactor the upper internal assembly must be removed to expose the fuel assemblies. With all the fuel assemblies removed the lower internal assembly can also be removed to facilitate inspection of itself and the reactor vessel.

The lower internal assembly basically consists of the cylindrical core barrel, with a support flange at its top and a massive lower core support structure welded to the bottom end of the barrel. All of the remaining components which in total form the lower internal assembly are attached to these structures and the complete assembly is supported by the flange at the top of the core barrel.

The remaining principal components of the NSSS are the reactor coolant pumps (RCP), the steam generators (SG) and the pressuriser.

9.3 Regions of Highest Radiological Hazard

9.3.1 As with the AGR, when considering PWR decommissioning, there are three major sources of radiological hazard, namely, radioactive source materials, activated materials and contaminated materials. In the PWR these hazards can be identified as follows:

(a) Radioactive Source Materials

These consist of the fuel assemblies which must be removed from the reactor via the existing fuel handling route before any decommissioning work can begin, after the reactor has been brought to a normal cold shutdown condition.

(b) Activated Materials

These consist of materials which have been in the path of the neutron stream emanating from the reactor core during the normal operational life of the reactor. These materials absorb the neutrons into their bulk, become activated, and in so doing emit other types of radiation, gamma radiation in particular. Materials activated in this way cannot be "de-activated" other than by time decay processes. The degree to which a structure becomes activated depends upon the material composition, the intensity of the neutron flux, and the distance of the structure from the reactor core. Hence the only significantly activated materials in the whole power station are:

1) The RPV shell and closure head package.

2) The RPV internal structures which support the core material and control rod mechanisms.

3) The region of the primary shield wall adjacent to the active core. The activated zone would extend approximately 1.0m to 1.5m radially into the wall from the reactor cavity face and similar dimensions above and below the active core region. See figures 48 to 52.

A comparison of activity levels in the above items has been compiled by Gregory[13] for 10 year and 100 year decay times. Total activity levels and mean specific activity levels have been extracted and are shown in table 2. Inspection of these values shows that the primary shield wall has significantly lower values of both total and mean specific activity levels than either the RPV or reactor internal structures. No other plant or structures become activated during normal operation of the plant.

(c) Contaminated Materials

Contamination consists generally of a surface deposit of radioactive material. Providing the contaminated surface is not generally porous, then the deposit can be removed by a washing or scouring process. The highest levels of contamination exist throughout the primary circuit, but many other secondary systems connected to the primary circuit will also be contaminated to a lesser degree during normal operation of the station. Generally, partial decontamination may be carried out during routine operation of the station but the reagents used are limited in strength to avoid damaging the surface of the pipework. However, during decommissioning the pipework systems can be thoroughly cleaned using more stringent methods as the pipework is not required for re-use. The effluent washings can be routed to the existing radioactive waste processing plant, the contaminants filtered out and disposed of safely.

Any areas of building structure which enclose activated systems, as outlined in section 9.2, may also be surface contaminated from liquid spillage or leakage and require treatment in a similar manner. High risk areas for contamination during normal operation or decommissioning should have suitable surface treatment to prevent absorption of spilt fluids.

9.3.2 From the above, if it is accepted that adequate decontamination of plant, pipework and structures can be achieved and that the fuel elements can be removed via normal handling routes, then it is the activated material which poses the greatest radiological hazard to be overcome during the decommissioning process.

9.4 Decommissioning Scenarios

9.4.1 Decommissioning of a PWR nuclear plant is as defined in Section 1.2 of this report.

9.4.2 In the following sections the term "dismantlement" is used to describe the organised, controlled and programmed sequence of events that constitutes the decommissioning process. These events include, but are not limited to, radiological protection requirements, plant and equipment removal and demolition of structures. For more detailed descriptions of these processes reference should be made to NUREG CR 0130[14].

9.4.3 Within the terms defined in Section 1.2 of this report, two basic scenarios exist for decommissioning of a PWR.

(a) Immediate Dismantlement

Radioactive source and activated materials are removed and the station is decontaminated and dismantled during the immediate period following cessation of power production. Upon completion of dismantlement the property may be released for unrestricted use.

(b) Safe Storage with Deferred Dismantlement

Radioactive source materials are removed, but activated and contaminated areas are secured to allow radioactivity to decay over a period to significantly lower levels. Only those structures and equipment necessary to ensure protection of the public from residual radioactivity are maintained, all other structures being removed. During this period of "safe storage" access to the facility remains under the control of a nuclear license. Dismantlement of the storage structures is deferred until radioactivity levels have decayed significantly. Upon completion of dismantlement the property may be released for unrestricted use.

9.4.4 The following orders of timescale apply to the above scenarios.

(a) Immediate dismantlement could take place within the first five to ten years following cessation of power production.

(b) Safe storage with deferred dismantlement would typcially mean a secure storage period of between ten and one hundred years after cessation of power production, to allow activity levels to decay, followed by dismantlement of the structures remaining at that time.

9.4.5 Obviously there are many variations that could occur to be intermediate between the above scenarios and the choice of the decommissioning method depends upon many factors. These factors include total radiation exposure limits for decommissioning staff and the general public, including transportation of active materials from the site, cost of maintenance and surveillance of structures and plant during storage periods, development costs of any specialist handling or dismantling techniques required which are not currently available, and so on.

9.4.6 The major events occurring during the decommissioning of a PWR may be outlined as follows:

(a) The reactor would be brought to cold shutdown condition, as for a routine refuelling outage.

(b) All the fuel assemblies would be removed from the reactor via the normal handling route to the spent fuel storage pond in the fuel building and then off site.

(c) The NSSS and other pipework systems would be decontaminated. After decontamination, systems which are no longer required can be dismantled. The arisings from the decontamination process could be collected, processed and transported off site using existing waste management systems.

(d) The reactor vessel upper and lower internal structures, having been removed from the reactor vessel and placed in their existing storage positions within the flooded refuelling pond of the reactor building, would be cut into sections, remotely under water, then removed in special shielding casks. Gregory[13] states that the remote handling and cutting of such items is within current technology but may require some development for this particular application.

(e) The refuelling pond could now be drained and the remaining NSSS plant and pipework dismantled having already been decontaminated. These items could be removed via existing plant access routes as whole plant items, or could be cut and handled piece small in shielded casks as required. Provision exists in current designs to allow the removal of the steam generators as whole units by the unbolting of precast wall panels above operating floor level of the reactor building structures. However, portions of the pressuriser cell would require demolition if the pressuriser is to be removed in one piece. All plant items can be handled using the reactor building polar crane. The removal of steam generators has already been carried out on the Surry and Turkey Point plants in the USA and Obrigheim in Germany.

(f) The only remaining items of high activity are the RPV and portions of the primary shield wall structure. The RPV could be cut piece small by remote handling operations and shipped off site in shielded casks. This is likely to be done in the dry with adequate shielding and due care would need to be exercised to collect and control the swarf cuttings.

(g) Demolition of the primary shield wall is a task with problems similar to those in removing the inner region of a PCRV wall. Both have active regions of about 1 m thick containing steel items. For the PWR, the dust and debris created by operations such as cutting and the like could be contained within the existing reactor building shell, or by a further containment barrier erected within the reactor building. There remains, however, the problem of removing the active steel embedments, particularly the vessel supports, from the wall.

(h) Once the activated materials had been removed and the structures decontaminated, demolition of the reactor building and any other remaining structures could proceed. The prestressing tendons of the reactor building, being ungrouted, can be easily removed.

9.5 Existing Structural Features of a PWR which may aid Decommissioning

(a) Fuel removal procedures follow normal refuelling operations as carried out throughout the station lifetime.

(b) The highly active reactor internal structures can be readily removed from the RPV shell and stored underwater awaiting disposal following procedures used throughout the working life of the station for periodic inspection.

(c) The NSSS plant is comparatively small in volume and its internal surface is specially prepared to facilitate decontamination.

(d) The NSSS can be readily dismantled into individual plant items which can be handled by the existing reactor building polar crane.

(e) As described in paragraph 9.4.6 (e) all major NSSS plant items except the RPV can be removed as whole units and extracted from the reactor building via the existing plant access penetration.

(f) All active areas are located within the reactor building which is a leaktight structure providing radiological shielding and protection from external hazards.

(g) All surfaces liable to receive contamination during station normal operation are protected either by continuously welded, stainless steel liners or by applied high efficiency decontaminable coatings.

(h) All decontamination arisings can be routed to the existing radioactive waste processing plant which can extract and package the waste in a suitable form for safe disposal.

(i) Shielding provided to reduce normal operation exposure limits will fulfil a similar function during decommissioning of non-active areas.

(j) Use of unbonded prestressing tendons in the reactor building shell, together with existing inspection platforms for access to the anchorages, will aid in destressing operations.

(k) The inbuilt redundancy and strength of the major structures would allow a flexible approach to establishing alternative routes for plant and equipment removal.

(l) Provision and maintenance of the lifetime records of all plant and structures, covering design, construction and operating information, normally specified as a requirement under Quality Assurance programmes, will be available to aid future decisions during decommissioning.

9.6 Structural Features that might be introduced into Future PWR Stations to aid Decommissioning

9.6.1 Active Structures

It was stated in section 9.3 that the reactor cavity face of the primary shield wall structure is likely to be the only intensely activated civil structure in the whole of the PWR station. Problems are mainly concerned with the removal of the highly active steel embedments and reinforcement and control of the activated dust and debris which would be generated during demolition of this structure. Hence, if the design and construction could be modified either to reduce the volume and level of activated material, or to allow dismantlement with minimum dusting, then these measures would greatly aid in the overall decommissioning process.

Difficulties occur in trying to design for ease of demolition and yet fulfil the required structural functions for normal operation and postulated fault conditions. Typically, in postulated fault conditions the primary shield wall is subject to large reactions from the embedded anchorages for the reactor pressure vessel as a result of high transient pressures within the reactor cavity and inspection gallery.

However, it may be feasible to design a system, possibly of interlocked, post-tensioned, prestressed concrete blocks, which would perform the structural duties required and maintain adequate biological shielding, but would not require cutting during demolition. As with the PCRV planes of weakness proposals, lack of fit due to tolerances would need to be considered.

Alternatively, to reduce the amounts of activated materials, it may be possible to replace part of the inner portions of the existing shield wall structure with a series of "water tanks" containing borated water which can easily be drained. This scheme, shown in figures 53 and 54, could replace as much as the innermost 600 mm of concrete, without major detriment to the structural strength. Such a scheme would reduce the volume of activated material of the primary shield wall and may also reduce the level of activity of the remaining structures but would have no effect on the level of activity in the vessel support system. The material content of the tanks themselves and their resultant activity would need to be considered.

9.6.2 Contaminated Structures

To aid with decontamination, all surfaces likely to be exposed to contaminated fluids during normal operation or fault conditions should have a readily decontaminable finish. This would generally take the form of protective painting or in the more likely areas of contamination a stainless steel liner could be employed as already used in the refuelling and fuel storage ponds. Such surface preparations are essential to prevent ingress of contaminated fluid into the body of the concrete structures. This is particularly pertinent to the reactor building structures, which would be subjected to contaminated containment spray water during some postulated faults, and regions of the auxiliary building which may be prone to spillage of fluids during routine operation or in post accident recirculation duty.

9.6.3 Demountable Structures

Greater consideration could be given during the design process to the removability of all plant and piping systems. This, combined with the greater use of unit construction techniques such as precast, bolted concrete elements, structural masonry or bolted structural steelwork connections, would allow better access to be established for plant and equipment removal.

9.6.4 Reduction in Pipework Support Points

A reduction in the total number of pipework support positions would reduce complexity and improve access in highly congested areas and would, of course, lead to simpler decommissioning of pipework systems.

10. REFERENCES

1. Council Decision of 27th March 1979 adopting a research programme concerning the decommissioning of nuclear power plants (79/344/Euratom). OJ No. L 83, 3.4.1979, p.19.

2. "The Community's Research and Development Programme on Decommissioning of Nuclear Power Plants - First Annual Progress Report (year 1980)", EUR 7440, 1981.

3. "The Community's Research and Development Programme on Decommissioning of Nuclear Power Plants - Second Annual Progress Report (year 1981)", EUR 8343, 1983.

4. "The Community's Research and Development Programme on Decommissioning of Nuclear Power Plants - Third Annual Progress Report (year 1982)", EUR 8962, 1984.

5. BOOTHBY, R.M. and WILLIAMS, T.M. The Control of Cobalt Content in Reactor Grade Steels. European Applied Research Reports Vol. 5, No. 2, 1983.

6. NEMET, J. and FRITZ, K. Concept, Construction, Testing and Operational Safety of a PCPV with Elastic Hot Liner and Adjustable Wall Temperature. Proc. of 5th International Conference on Structural Mechanics in Reactor Technology. Berlin, August 1979.

7. WOOLLAM, P.B. and PUGH, I.G. Neutron Induced Activation, Waste Disposal and Radiation Levels for the Reactor Island Structure of a Decommissioned Magnox Power Station. IAEA-SM-234/10.

8. McFARLAND, J.M. Demolition of Concrete Structures by Heat - A Preliminary Study. Proc. of the ANS Topical Meeting, 1979, pp457-469.

9. DYSON, J.A. and HARRISON, J.R. The Dependence of Fast Neutron Attenuation in Portland Cement on its Hydrogen Content. AERE, R/R 1942. April 1956.

10. HENRIE, J.O. Properties of Nuclear Shielding Concrete. Journal of the American Concrete Institute. July 1959.

11. ITO, N. and YANAGAWA, A. Method for Manufacturing High Boron Concrete for Radiation Protection. Japanese Patent No. 43-11793, May 1983.

12. BS 4975 : 1973, Specification for Prestressed Concrete Pressure Vessels for Nuclear Reactors, British Standards Institution.

13. GREGORY, A.R. CEGB Proof of Evidence on Decommissioning. Sizewell 'B' Power Station Public Enquiry. November 1982.

14. NUREG, CR 0130. Technology, Safety and Costs for Decommissioning of a Reference Pressurised Water Reactor. U.S. Nuclear Regulatory Commission. June 1978.

11. ACKNOWLEDGEMENTS

On behalf of Taylor Woodrow Construction Limited, the authors gratefully acknowledge the financial and technical support afforded to them during the study by the Commission of the European Communities, the Nuclear Installations Inspectorate, the Central Electricity Generating Board, the National Nuclear Corporation Limited and the Department of the Environment. Thanks is also given to the Central Electricity Generating Board for its permission to reproduce figures nos. 1, 2, 3, 4 and 8.

The authors also acknowledge the help given to them by their colleagues in carrying out the work and preparing the report.

12. TABLES AND FIGURES

TABLE 1 : RESULTS OF ULTIMATE LOAD ANALYSES OF STRUCTURES HAVING PLANES OF WEAKNESS

STRUCTURE	VESSEL PRESSURE AT ULTIMATE LOAD (p.s.i)	LOAD FACTOR	FAILURE CRITERIA
Type 1 (No planes of weakness)	1632	2.534	Longitudinal crack width
Type 2, with all reinforcement outside planes of weakness zone	1624	2.521	Longitudinal crack width
Type 2, with haunch reinforcement increased by 20%	1631	2.533	Longitudinal crack width

TABLE 2 : COMPARISON OF ACTIVITY LEVELS IN PWR FOR 10 YEAR AND 100 YEAR DECAY TIMES

(Extracted from Ref. 11)

COMPONENT	MATERIAL	MASS (t)	10 YEAR DECAY		100 YEAR DECAY	
			TOTAL ACTIVITY (Ci)	MEAN SPECIFIC ACTIVITY (Ci t^{-1})	TOTAL ACTIVITY (Ci)	MEAN SPECIFIC ACTIVITY (Ci t^{-1})
Activated Plant:						
reactor internals	stainless steel	210	840000	4000	110000	520
reactor pressure vessel	mild steel	400	1700	4.3	40	0.1
biological shield	concrete and reinforcement	600	80	0.13	4	0.01
Contaminated Plant:						
- in containment:						
steam generators	stainless steel	1250	700	0.56	100	0.08
primary coolant circuit	stainless steel	520	37	0.07	5	0.009
- in other buildings:						
auxiliary building	steel	145	36	0.25	-	-
fuel building	steel	12	3	0.25	-	-
radwaste building	steel	80	25	0.31	-	-
Contaminated Structures:						
- in containment	concrete	1900*	0.2**	0.0001	0.03**	0.000016
- in containment	mild steel	670*	0.6**	0.0009	0.08**	0.00012
- in other buildings	concrete	5700*	0.8**	0.00014	-	-
- in other buildings	steel	1100	1.4**	0.0013	-	-

Notes:

1 * These data refer to the maximum values of the estimated quantity of contaminated concrete surfaces. The minimum quantity may be a factor 50 less.

2 ** These data refer to the maximum values of the estimated activity levels. The minimum levels may be a factor 300 less.

3 No credit is taken in this data for any decontamination operations which may precede decommissioning.

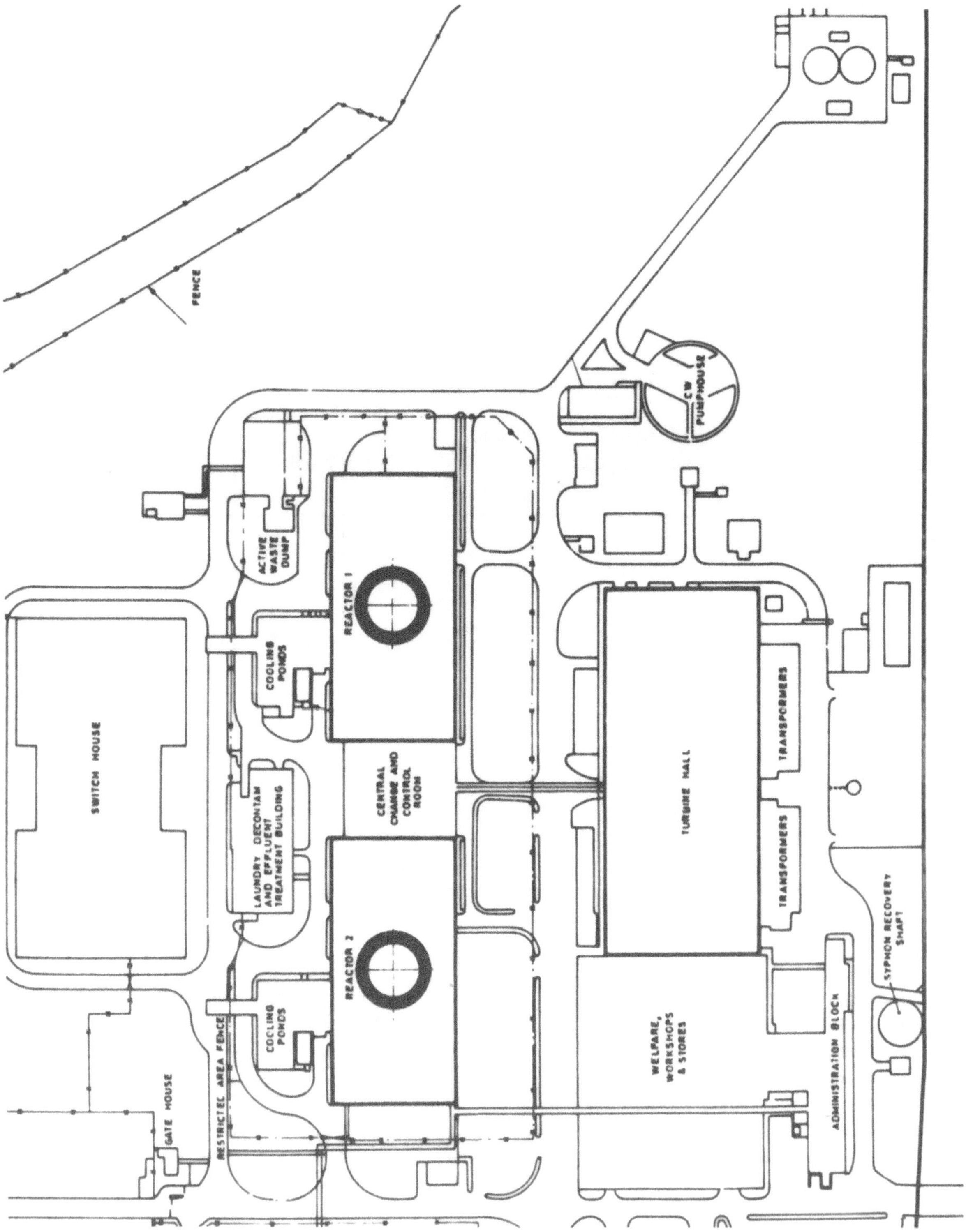

Fig 1. SITE PLAN–TYPICAL MAGNOX STATION

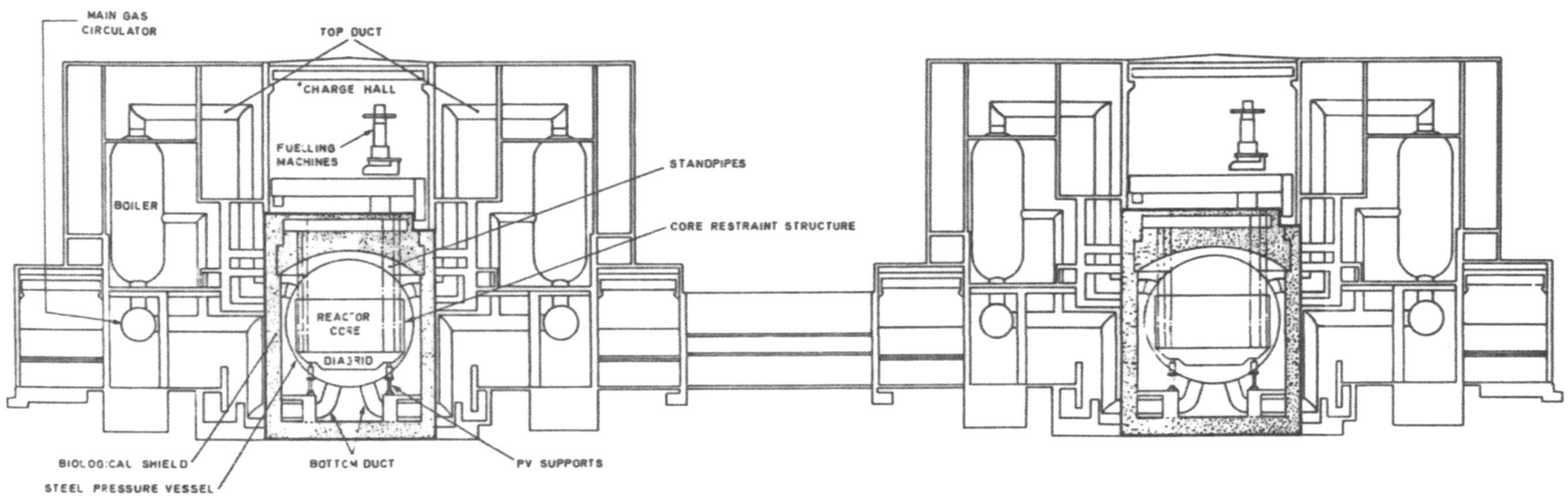

Fig 2. SECTION THROUGH TYPICAL MAGNOX STATION

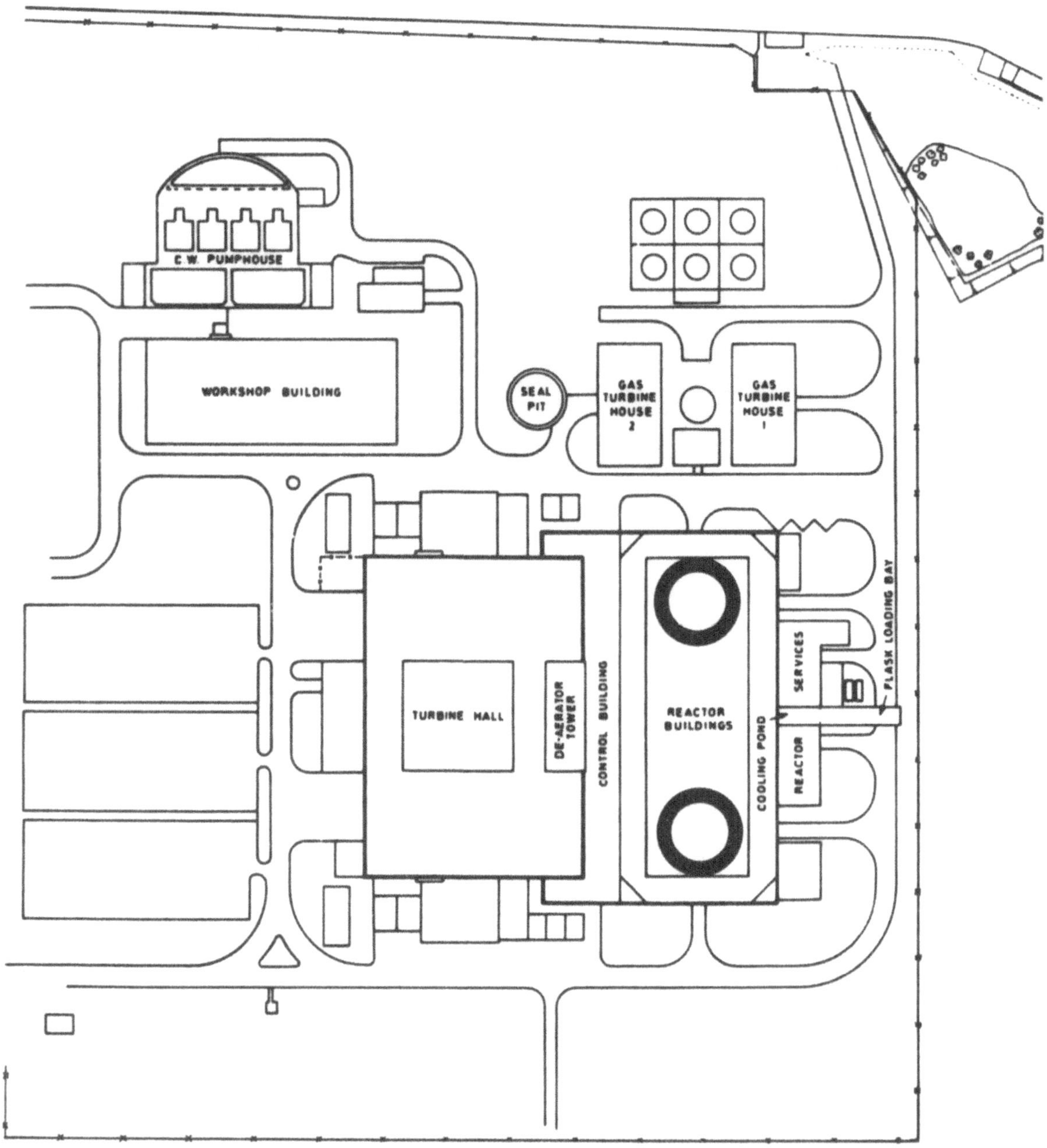

Fig 3. SITE PLAN–TYPICAL AGR STATION

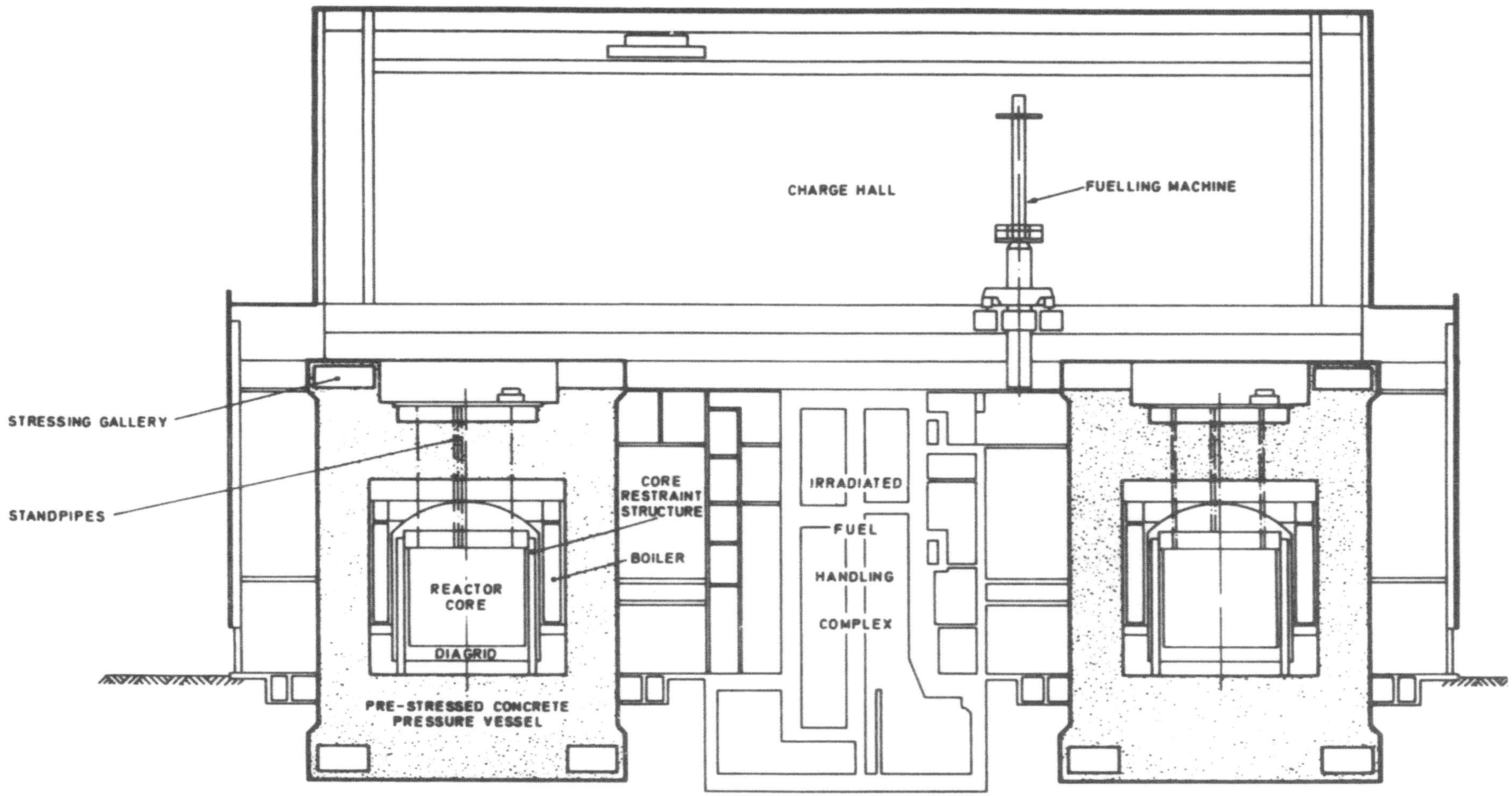

Fig 4. SECTION THROUGH TYPICAL AGR STATION

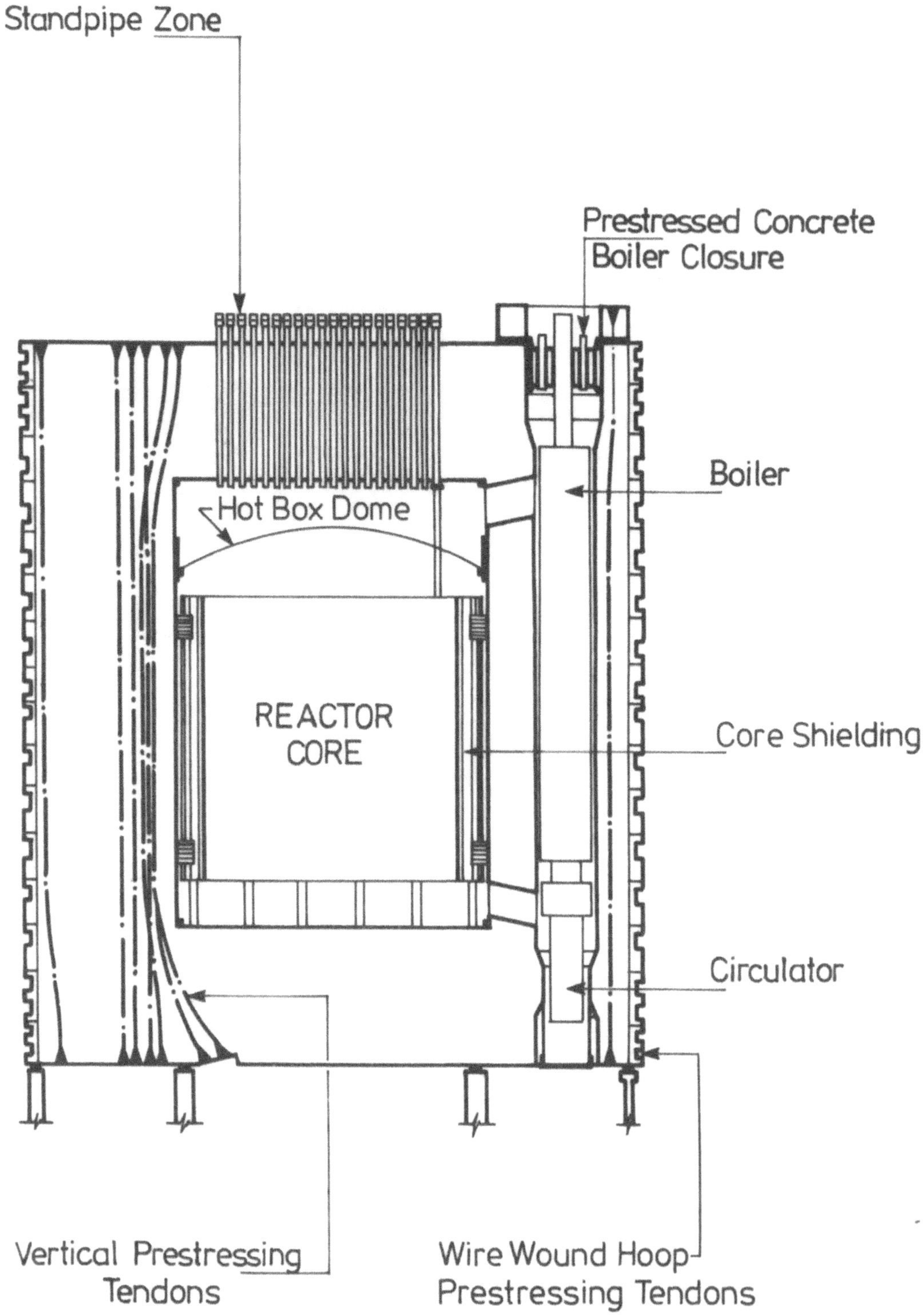

Fig. 5. SECTION THROUGH HARTLEPOOL/HEYSHAM 1 MULTICAVITY REACTOR VESSEL

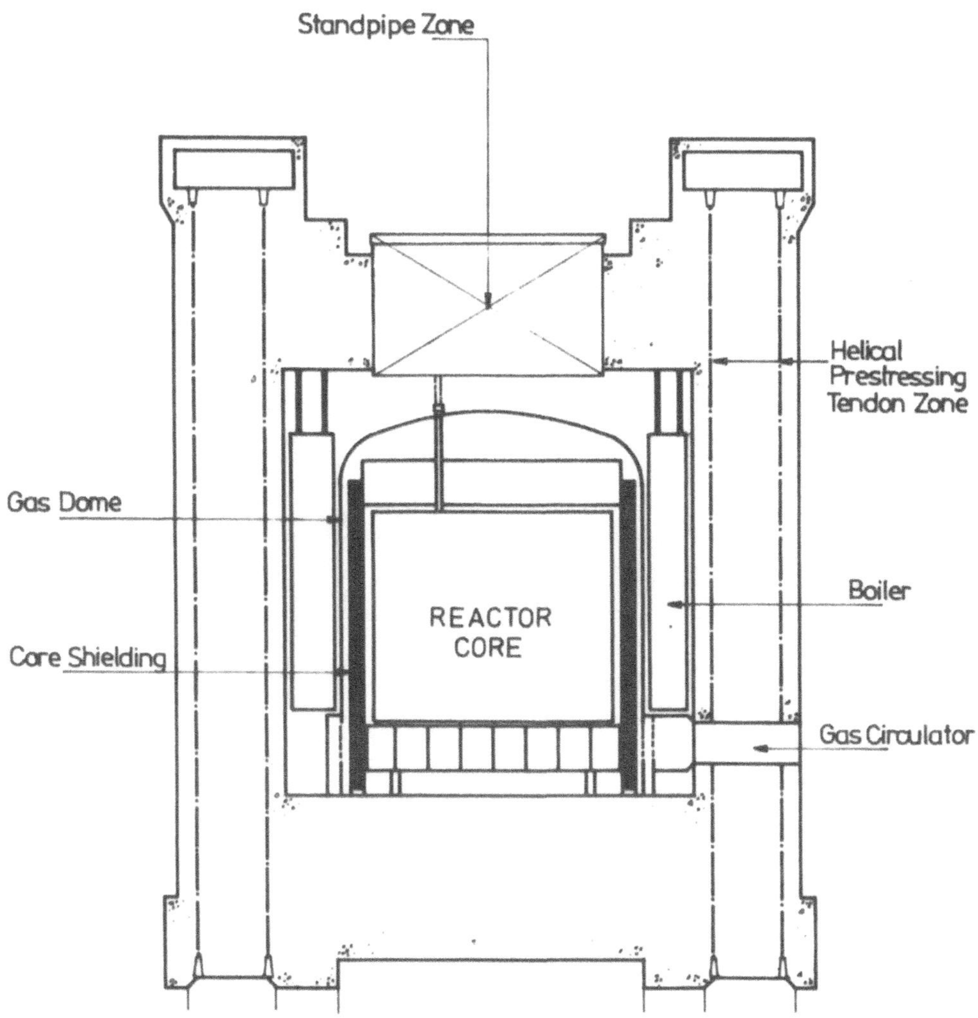

Fig 6 SECTION THROUGH HEYSHAM II/ TORNESS SINGLE CAVITY REACTOR VESSEL

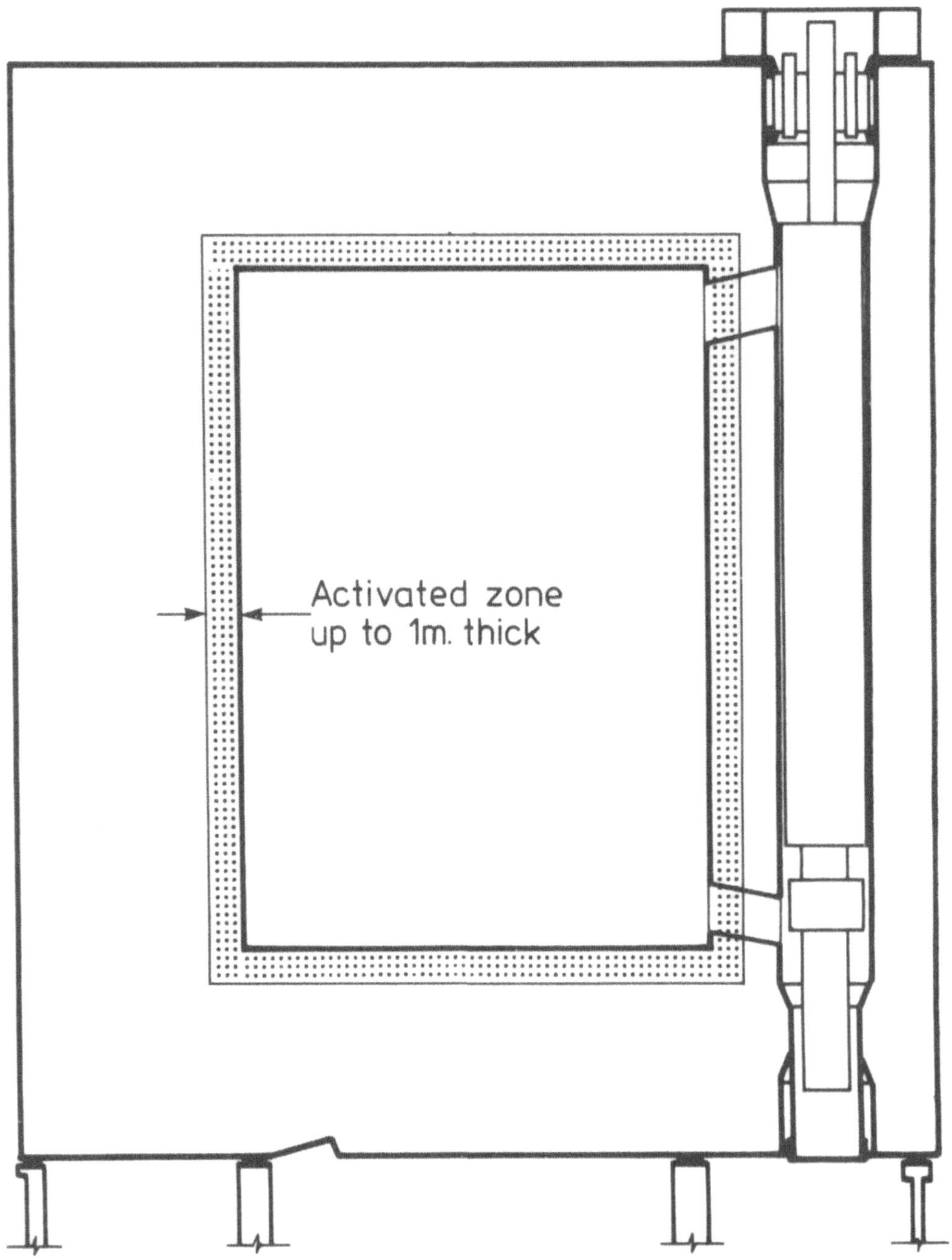

Fig 7. ASSUMED ACTIVATED ZONE OF A TYPICAL A.G.R. PRESTRESSED CONCRETE PRESSURE VESSEL

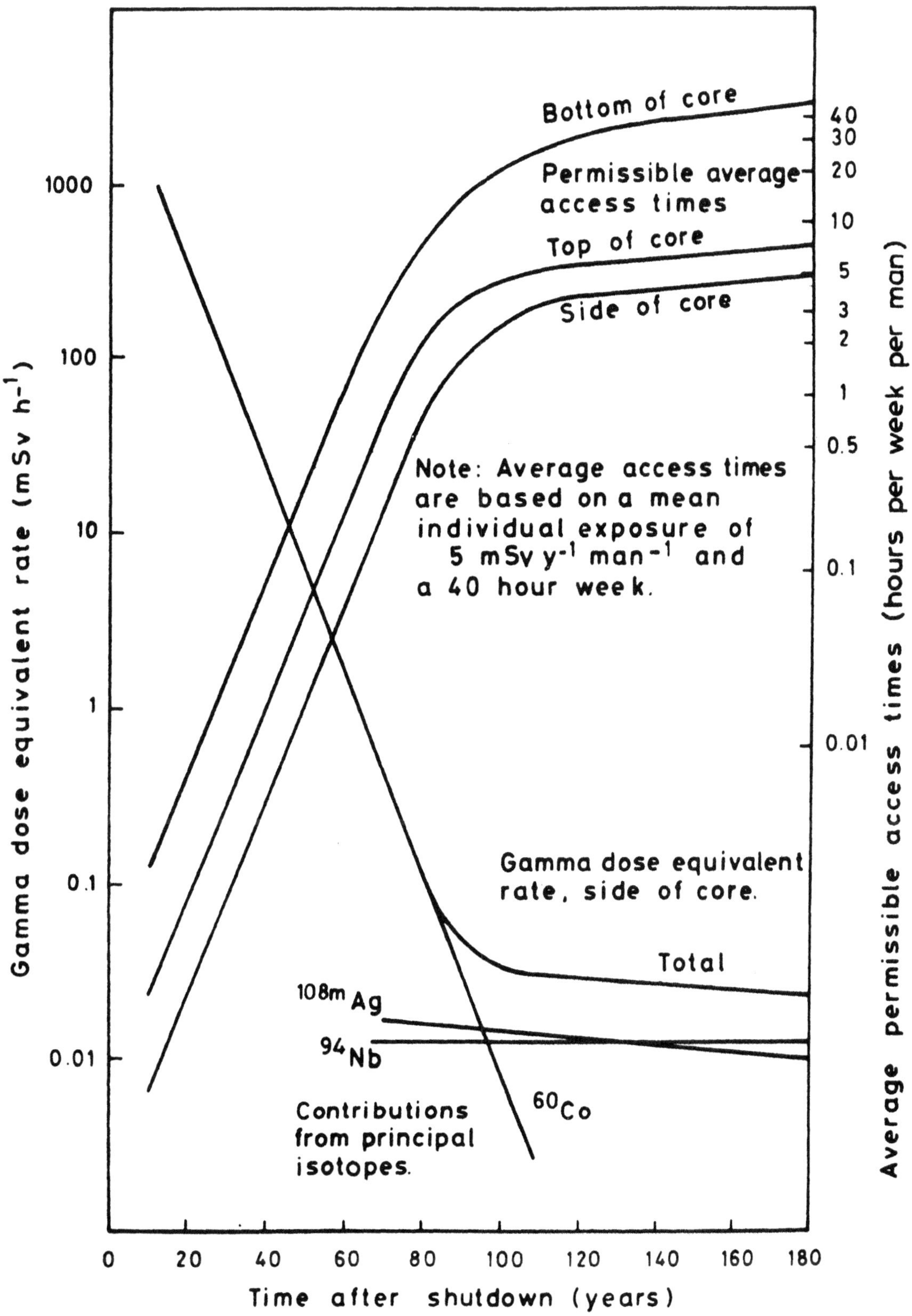

Fig 8. VARIATION OF AVERAGE PERMISSIBLE ACCESS TIMES INSIDE VESSEL (WITH CORE INTACT) AND DOSE EQUIVALENT RATE, WITH TIME AFTER SHUTDOWN.

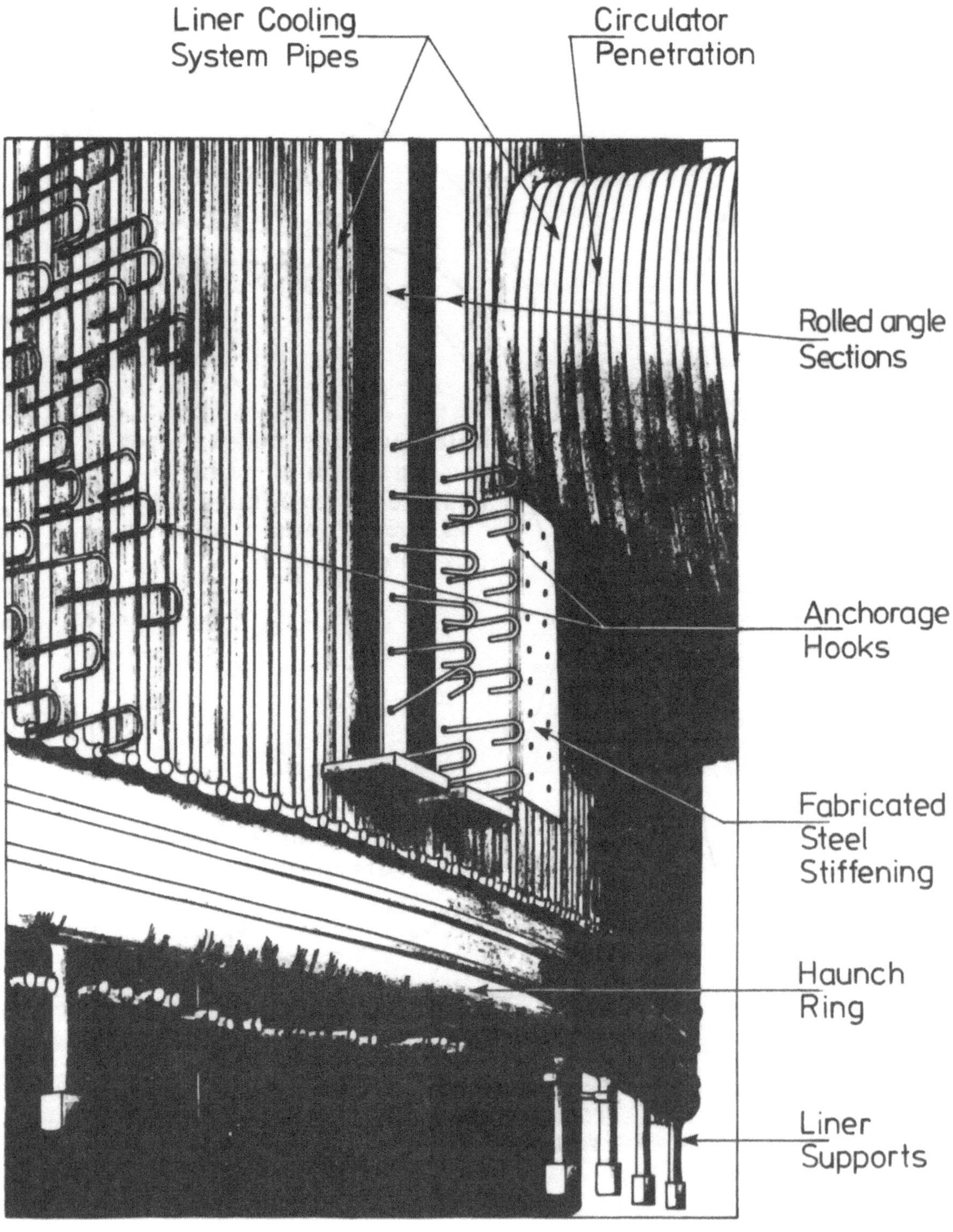

Fig 9. VIEW ON OUTSIDE FACE OF A TYPICAL PCRV LINER

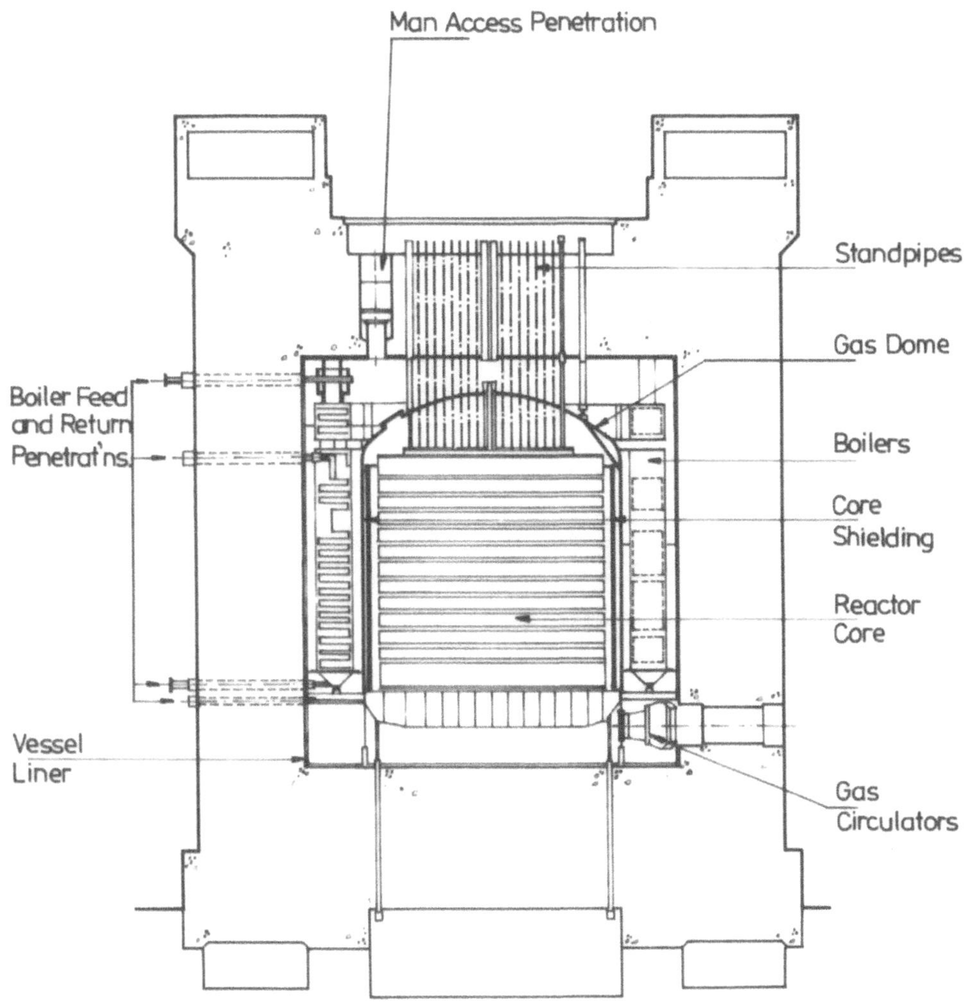

Fig.10. SECTION THROUGH TYPICAL PCRV SHOWING MAIN INTERNAL STRUCTURES AND ITEMS

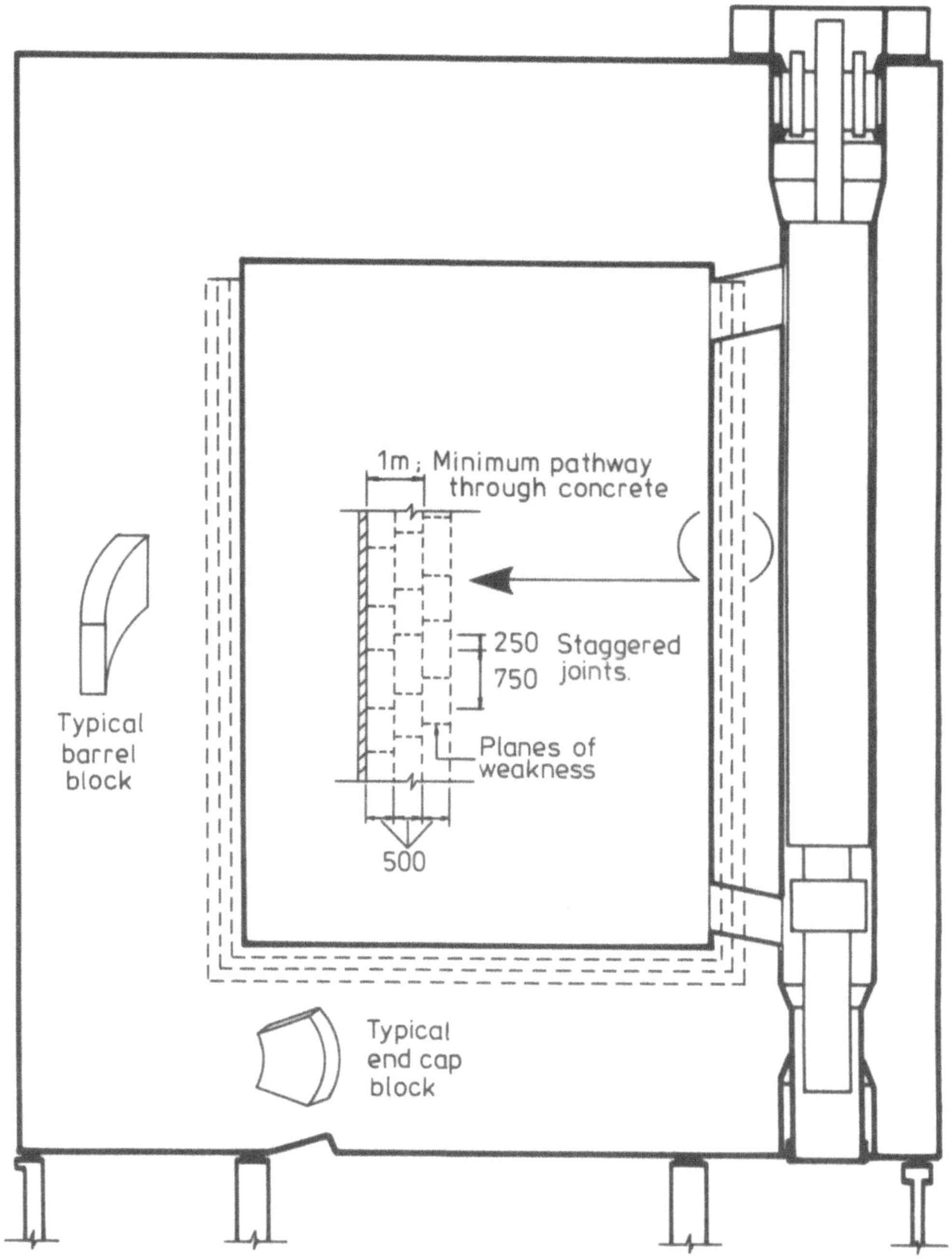

Fig 11. PROPOSED ARRANGEMENT OF PLANES OF WEAKNESS IN THE ACTIVATED ZONE OF A PRESTRESSED CONCRETE PRESSURE VESSEL

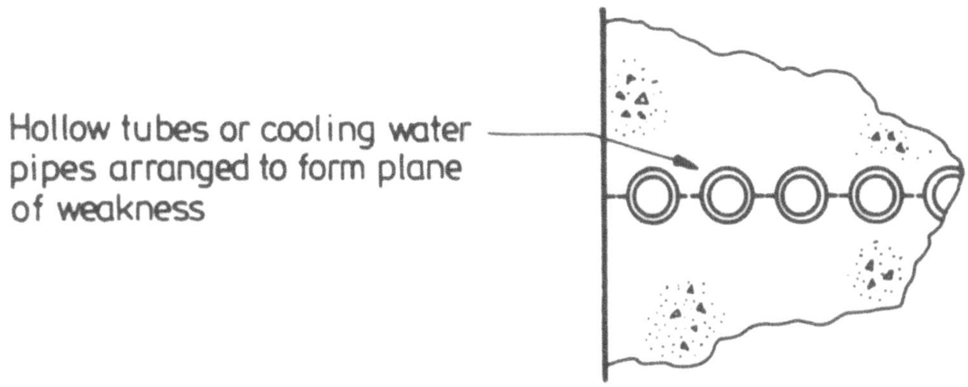

Fig.12. PLANES OF HOLLOW TUBES

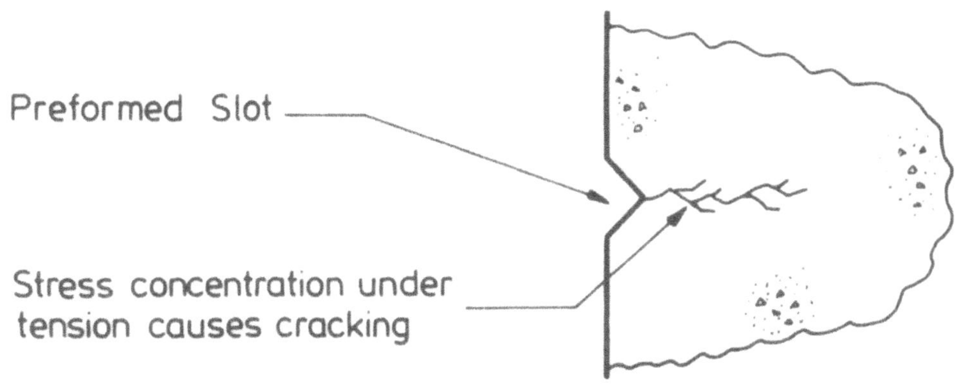

Fig.13. CRACK INDUCER SLOTS

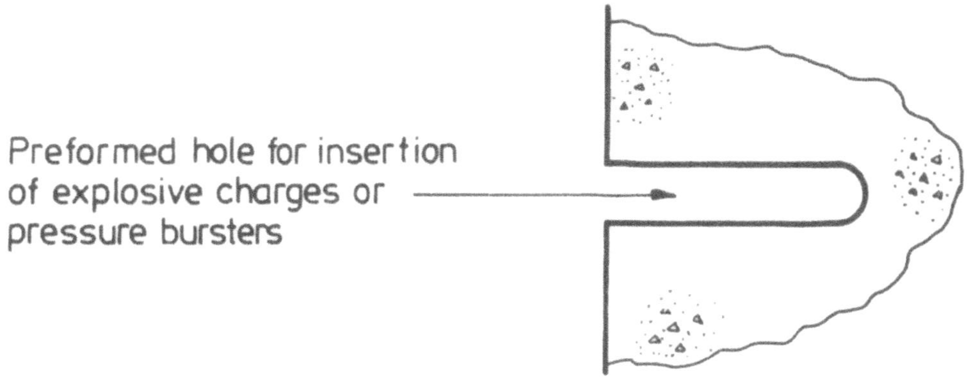

Fig.14. PREFORMED CAVITIES

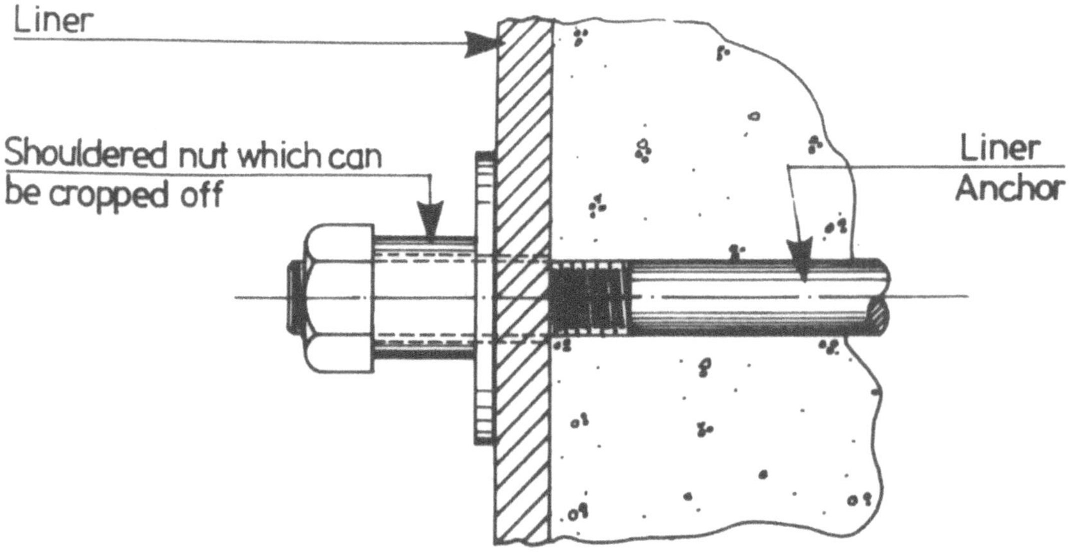

Fig.15. ALTERNATIVE LINER ANCHOR DETAIL (a)

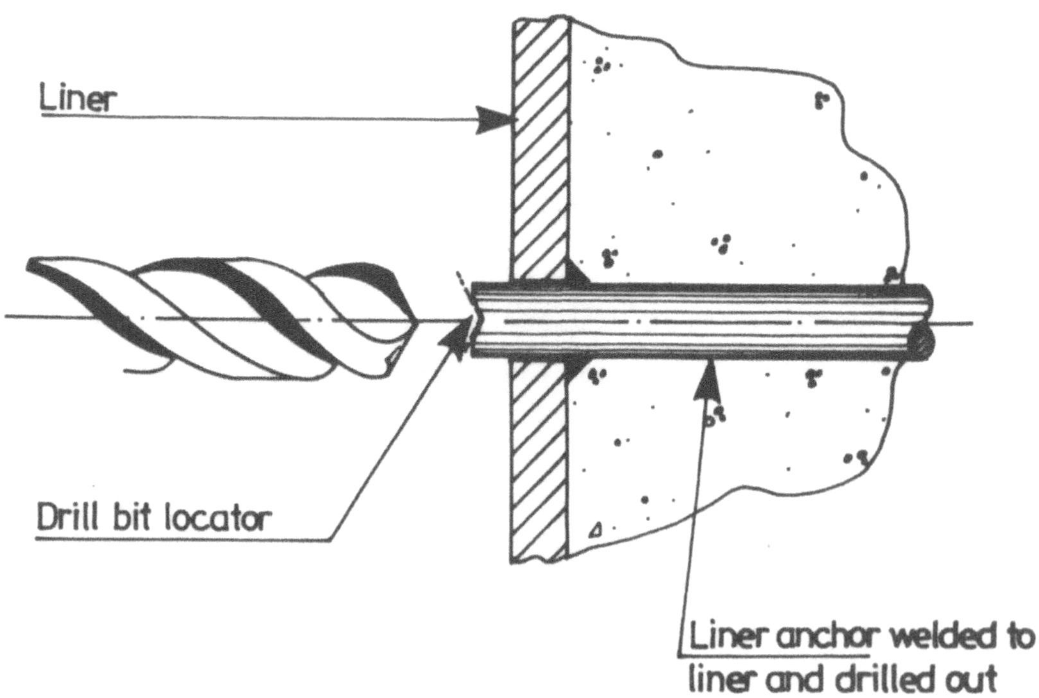

Fig.16. ALTERNATIVE LINER ANCHOR DETAIL (b)

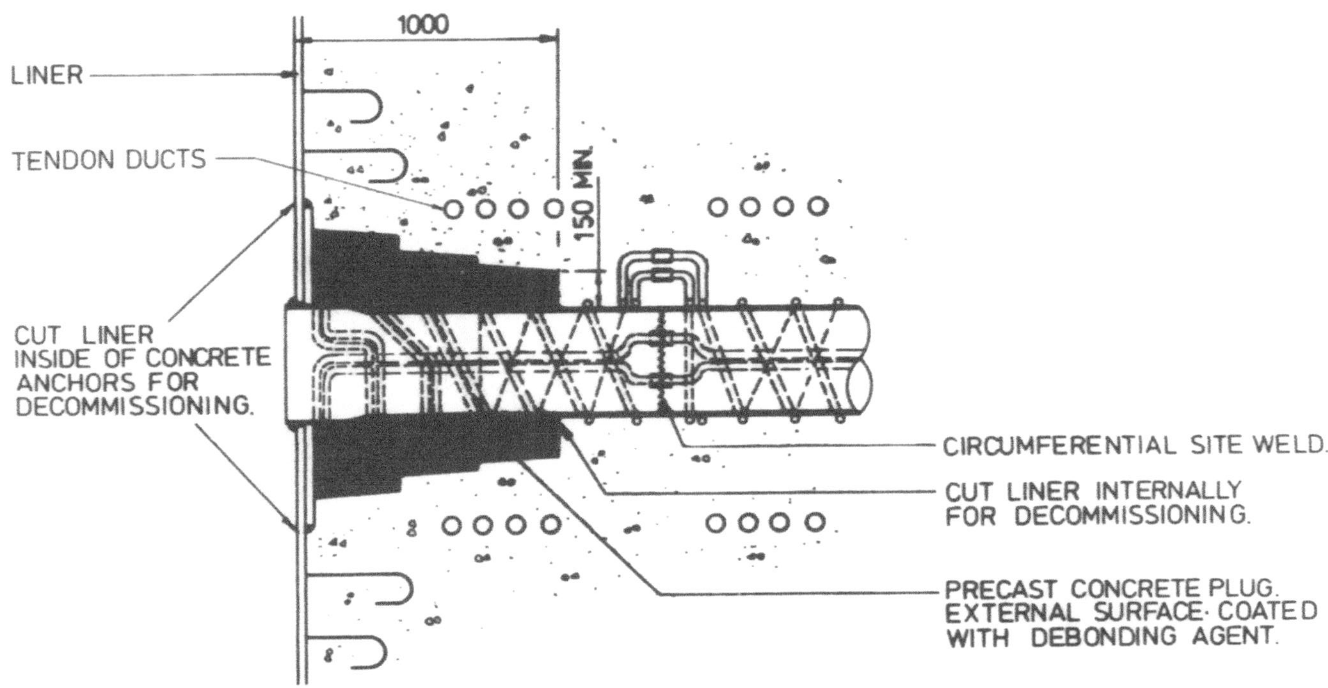

Fig. 17. **POSSIBLE DETAIL TO FACILITATE REMOVAL OF PENETRATION END ATTACHED TO MAIN LINER**

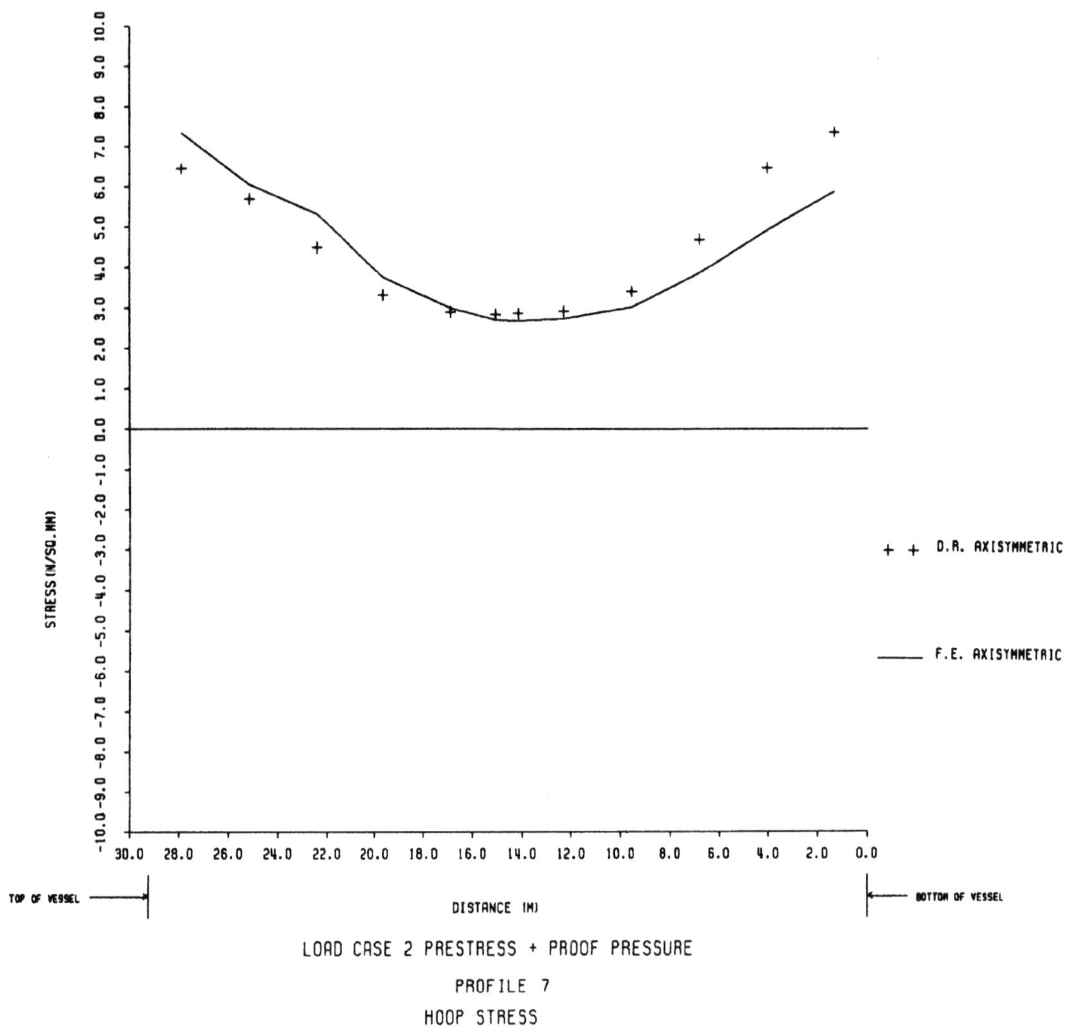

FIG. 25 PLANES OF WEAKNESS ANALYSIS
TYPICAL COMPARISON OF D.R. AND F.E. ANALYSES
OF STRUCTURE TYPE 1

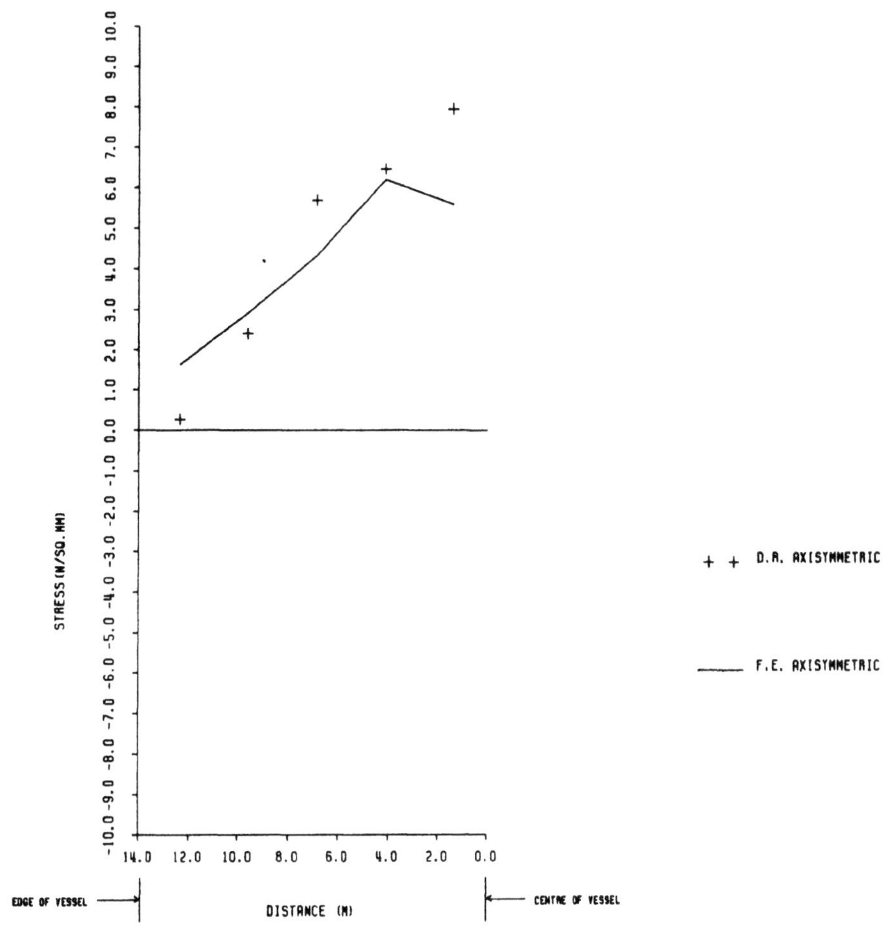

LOAD CASE 2 PRESTRESS + PROOF PRESSURE

PROFILE 2

RADIAL STRESS

FIG. 24 PLANES OF WEAKNESS ANALYSIS
TYPICAL COMPARISON OF D.R. AND F.E. ANALYSES
OF STRUCTURE TYPE 1

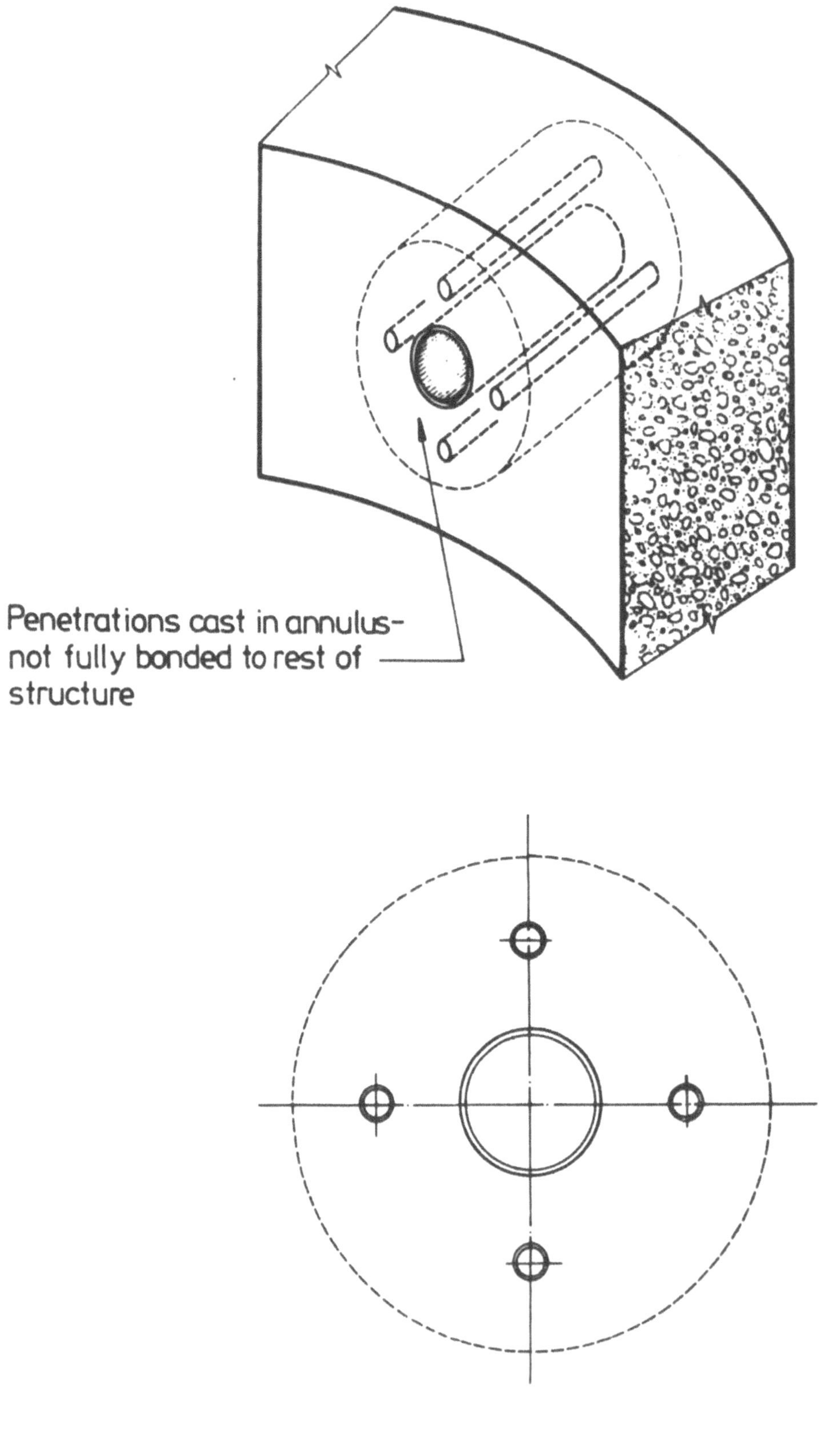

Fig.18. GROUPED AND DEBONDED PENETRATIONS

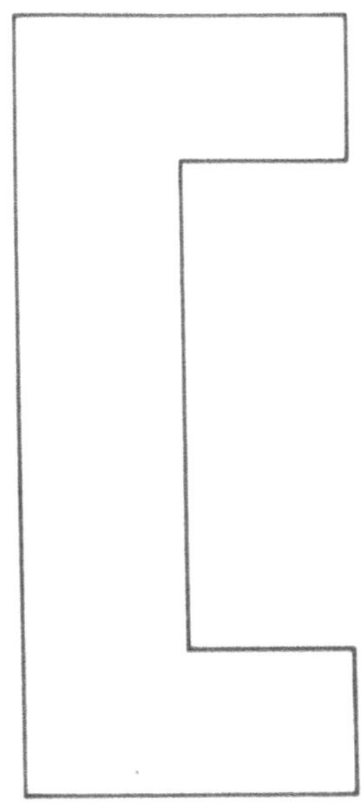

Fig. 19.
PLANES OF WEAKNESS ANALYSIS
STRUCTURE TYPE 1

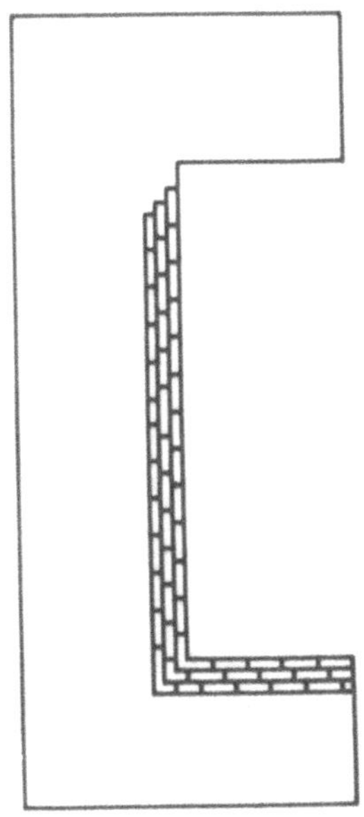

Fig. 20.
PLANES OF WEAKNESS ANALYSIS
STRUCTURE TYPE. 2

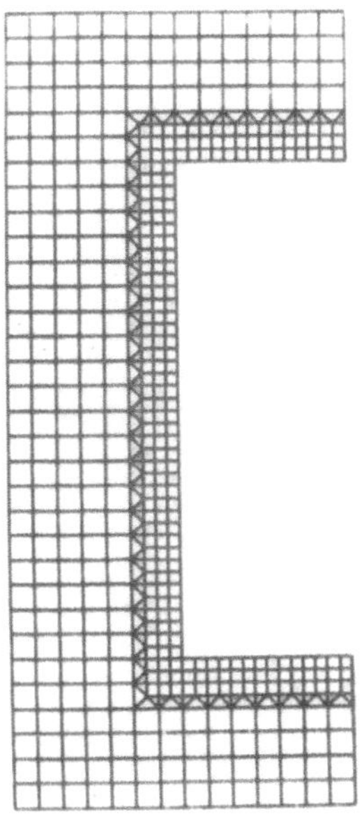

Fig. 21.
FINITE ELEMENT GRID FOR
ASSESSMENT OF STRUCTURAL EFFECTS
OF PLANES OF WEAKNESS

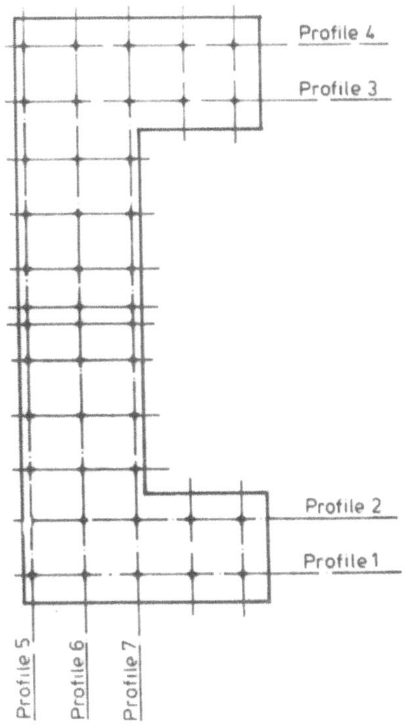

Fig. 22.
PLANES OF WEAKNESS ANALYSIS
LOCATION OF STRESS PROFILE
CONTROL POINTS

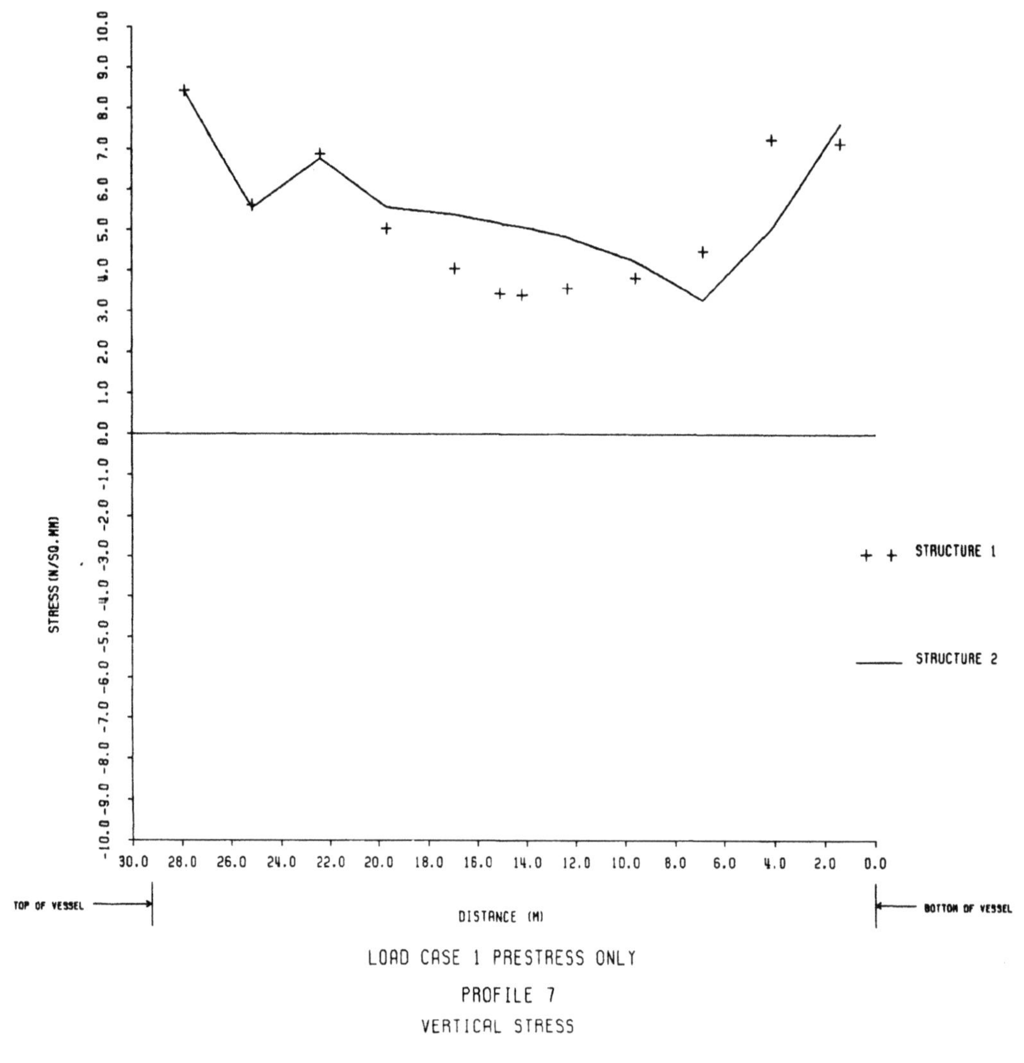

FIG. 27 PLANES OF WEAKNESS ANALYSIS
TYPICAL COMPARISON OF STRESS PROFILES FOR
STRUCTURES TYPE 1 AND 2

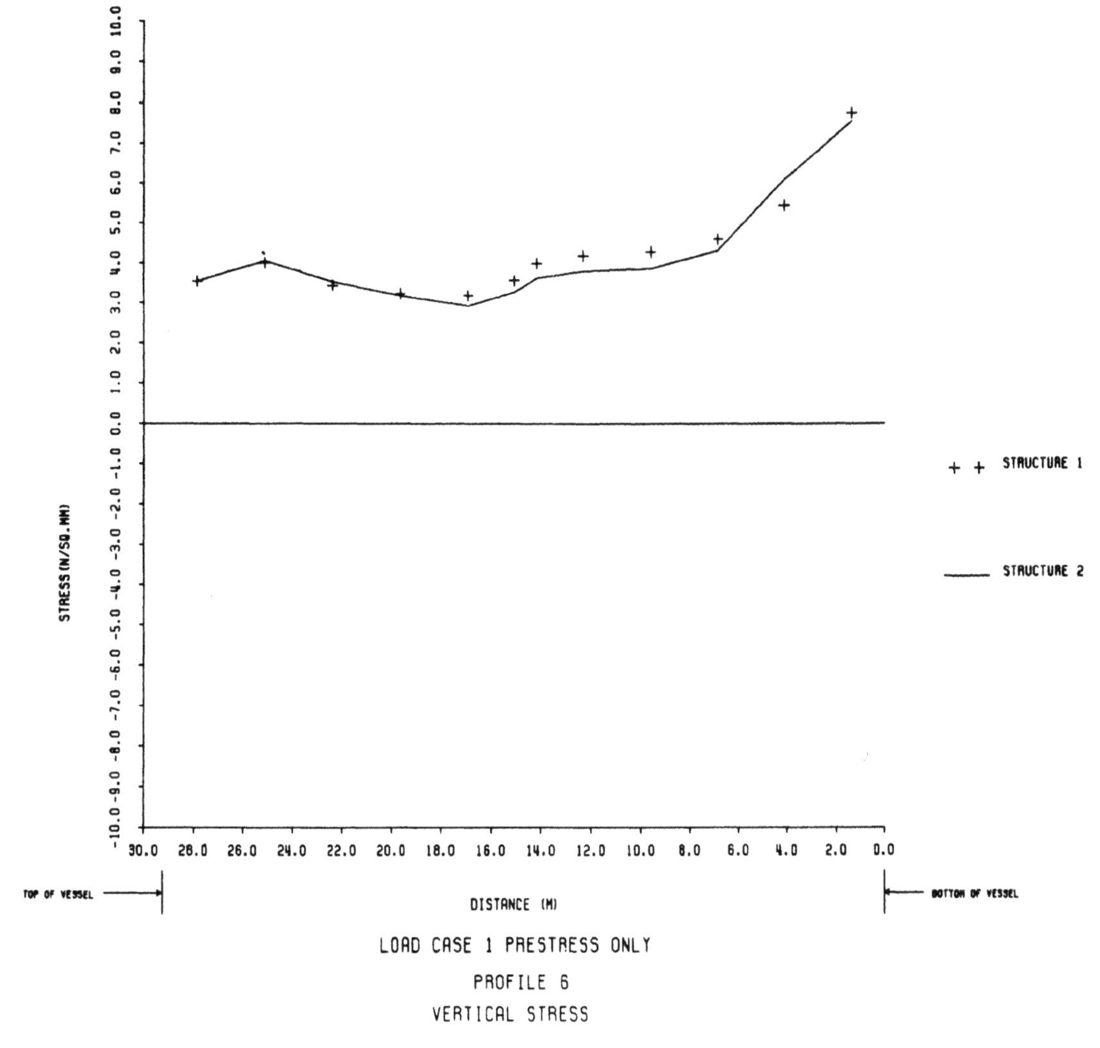

FIG. 26 PLANES OF WEAKNESS ANALYSIS
TYPICAL COMPARISON OF STRESS PROFILES FOR
STRUCTURES TYPE 1 AND 2

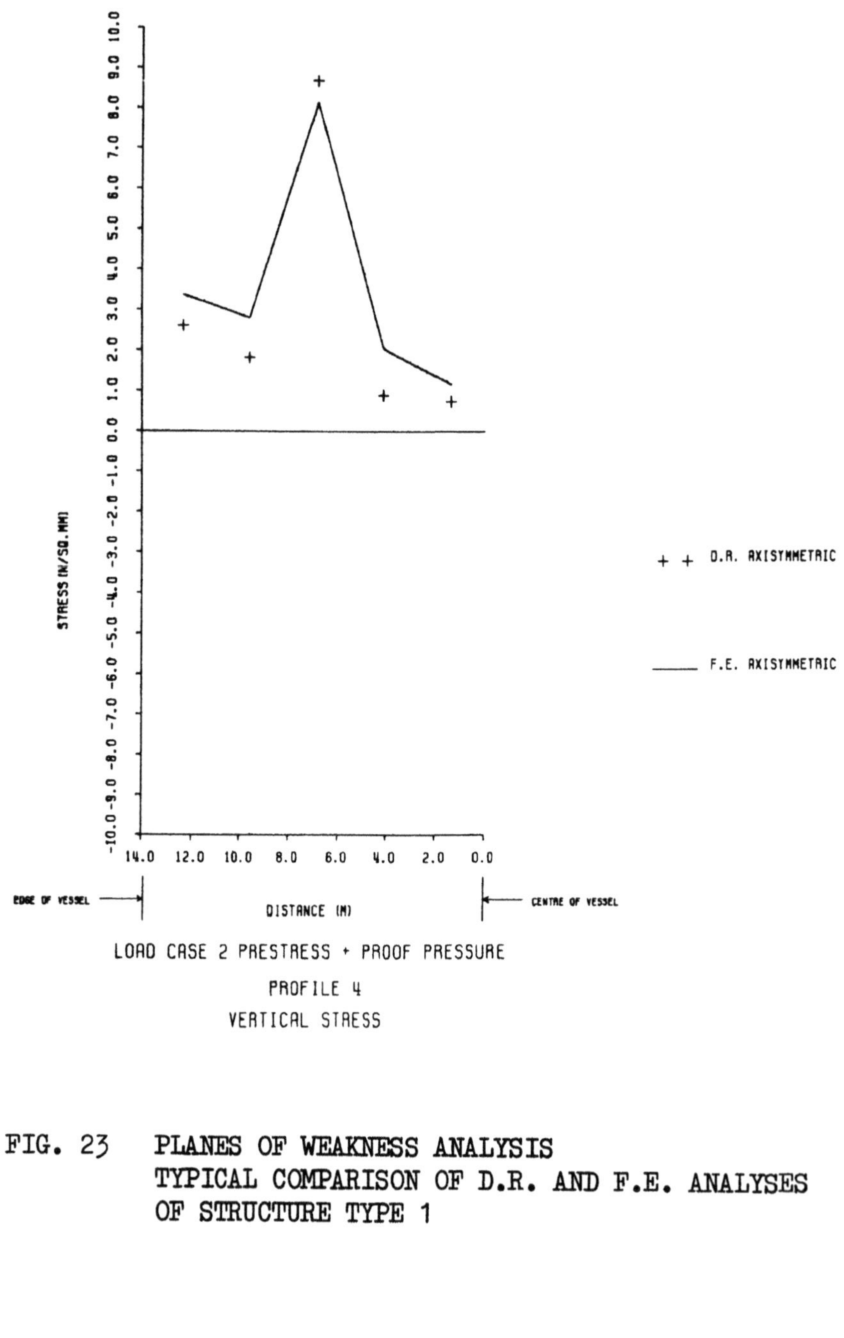

FIG. 23 PLANES OF WEAKNESS ANALYSIS
TYPICAL COMPARISON OF D.R. AND F.E. ANALYSES
OF STRUCTURE TYPE 1

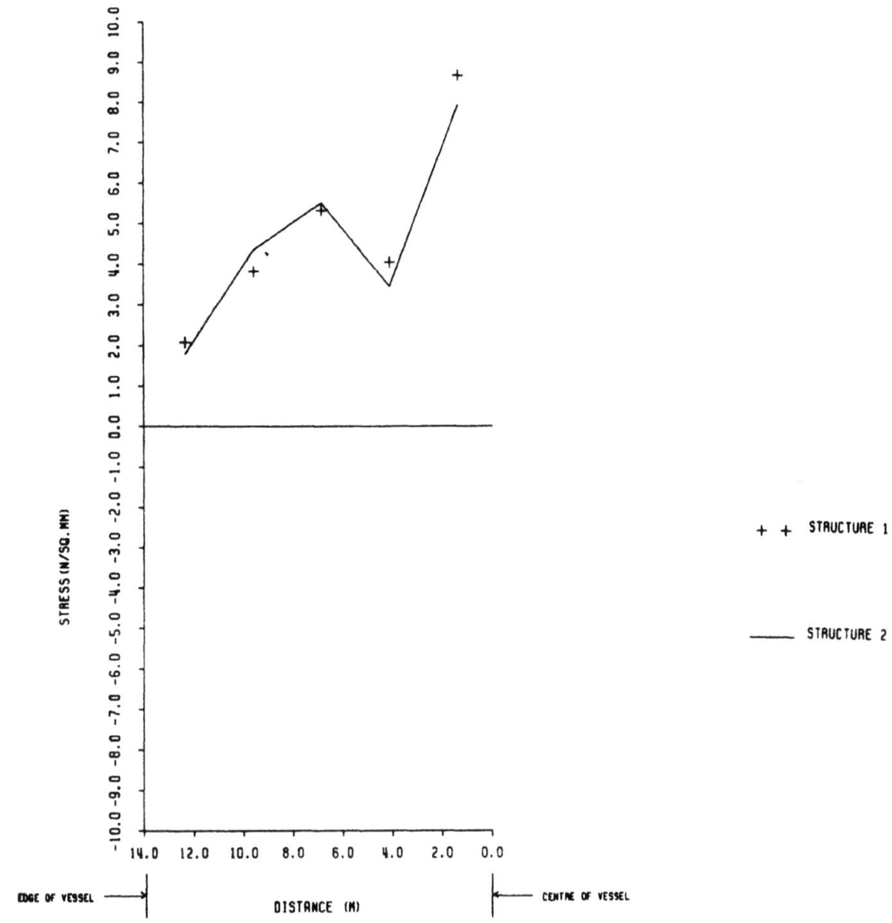

LOAD CASE 1 PRESTRESS ONLY

PROFILE 2

RADIAL STRESS

FIG. 28 **PLANES OF WEAKNESS ANALYSIS TYPICAL COMPARISON OF STRESS PROFILES FOR STRUCTURES TYPE 1 AND 2**

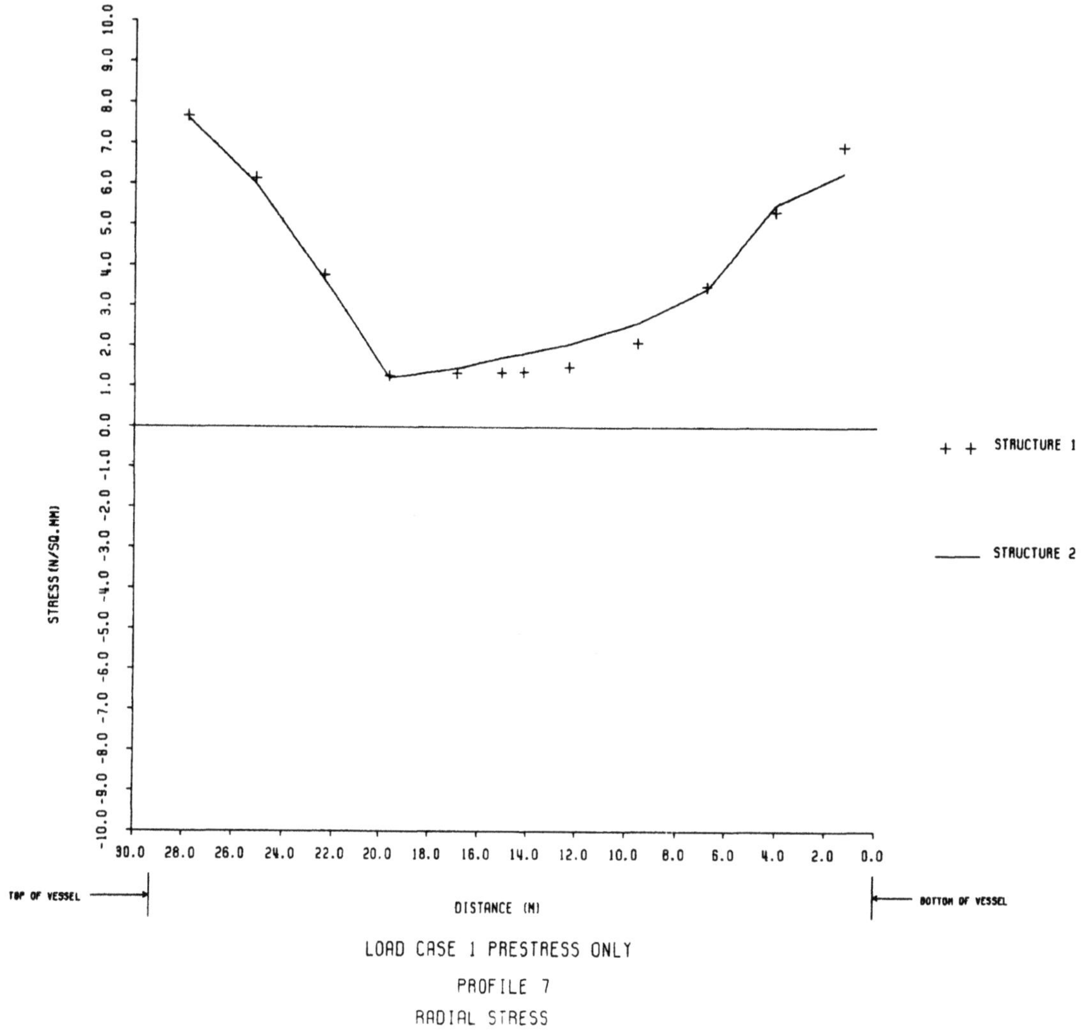

FIG. 29 PLANES OF WEAKNESS ANALYSIS
TYPICAL COMPARISON OF STRESS PROFILES FOR
STRUCTURES TYPE 1 AND 2

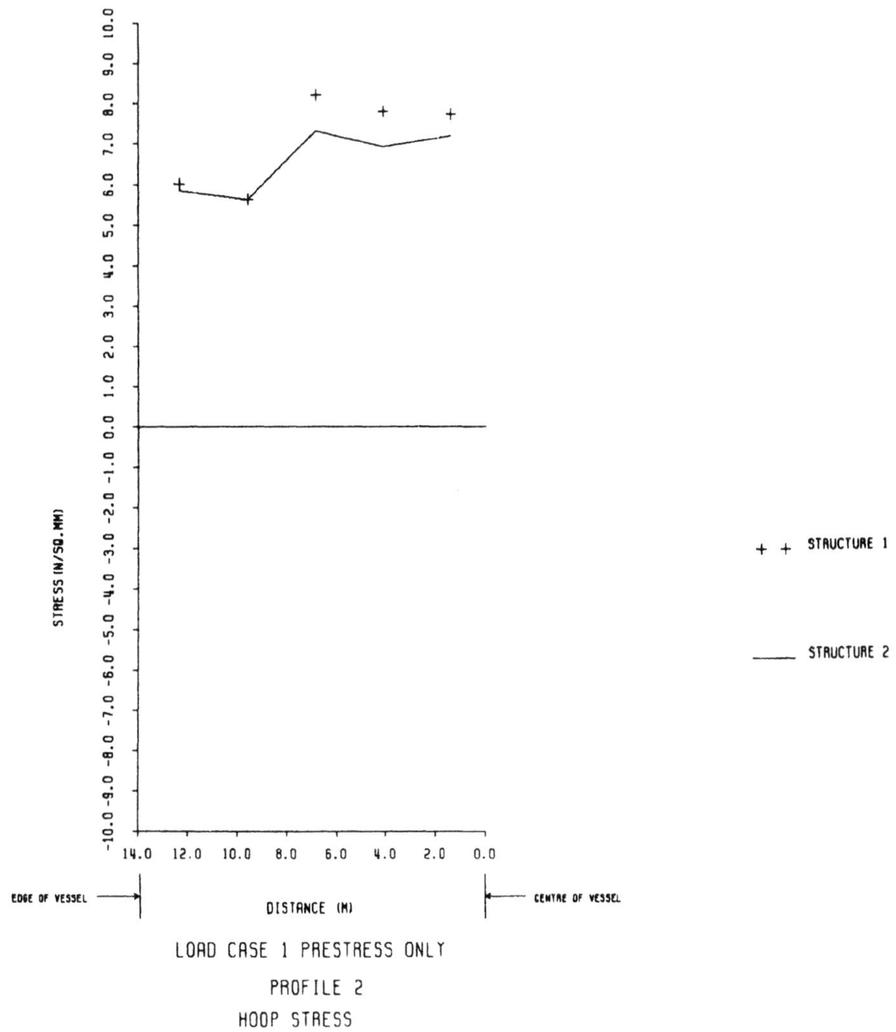

FIG. 30 PLANES OF WEAKNESS ANALYSIS
TYPICAL COMPARISON OF STRESS PROFILES FOR
STRUCTURES TYPE 1 AND 2

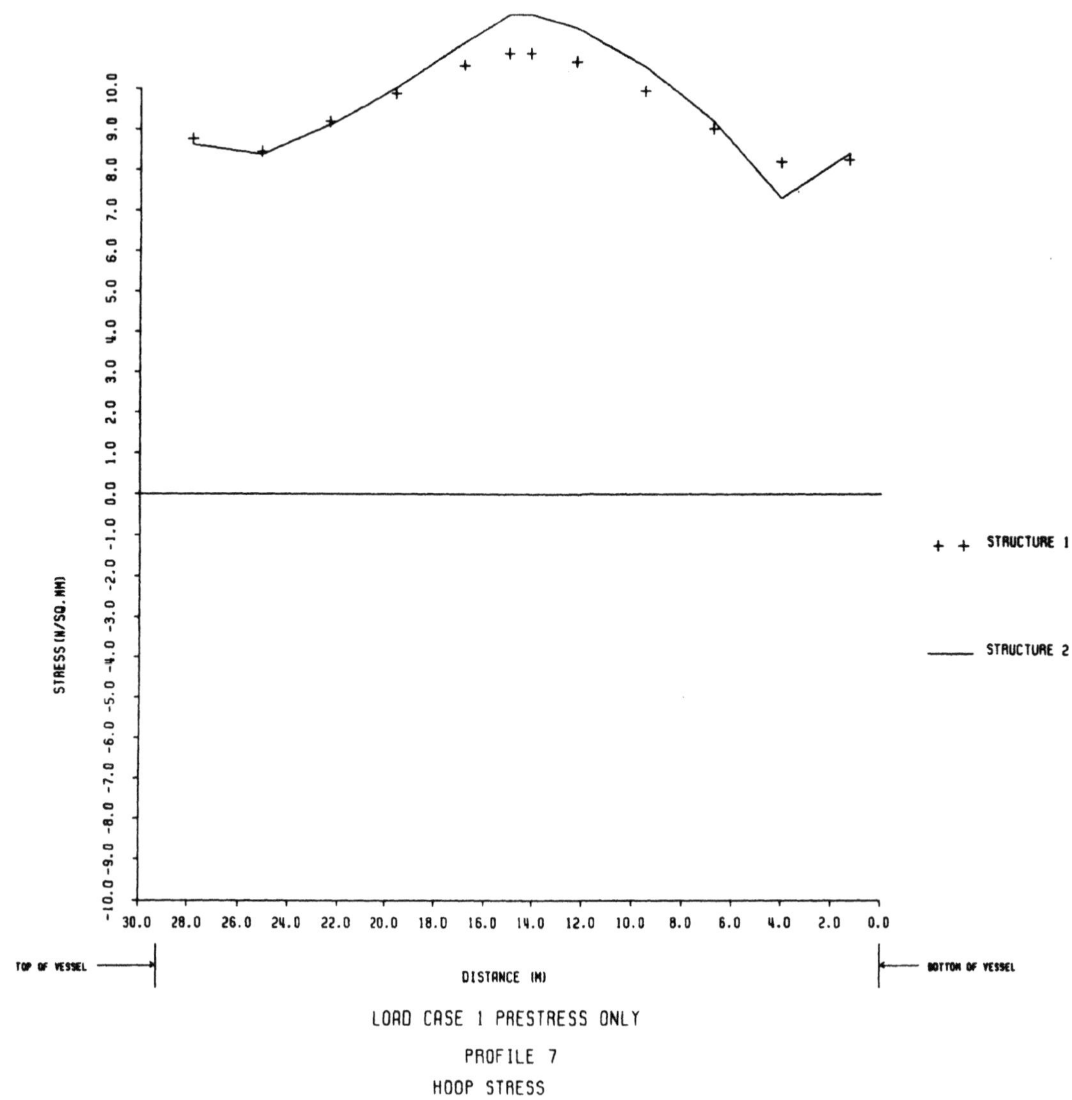

FIG. 31 PLANES OF WEAKNESS ANALYSIS
TYPICAL COMPARISON OF STRESS PROFILES FOR
STRUCTURES TYPE 1 AND 2

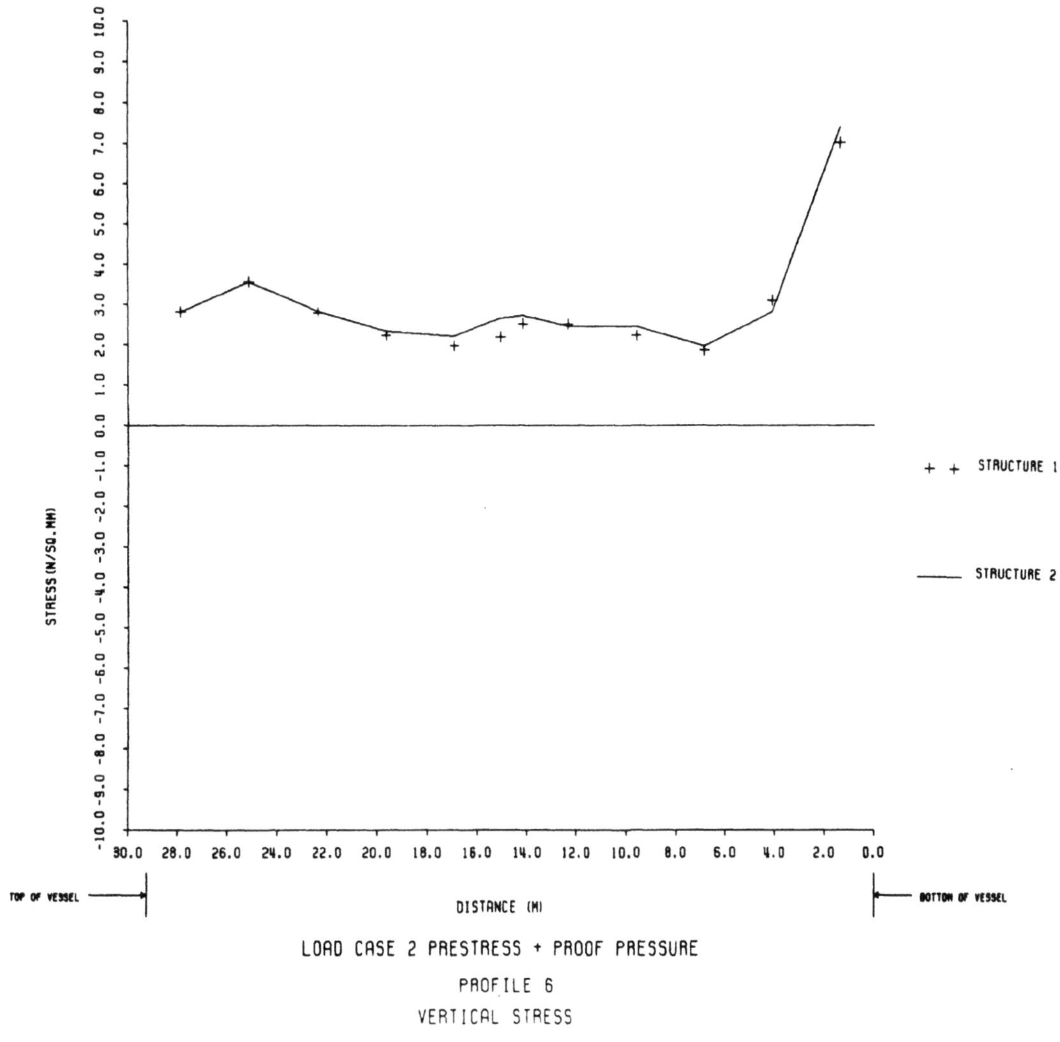

FIG. 32 PLANES OF WEAKNESS ANALYSIS
TYPICAL COMPARISON OF STRESS PROFILES FOR
STRUCTURES TYPE 1 AND 2

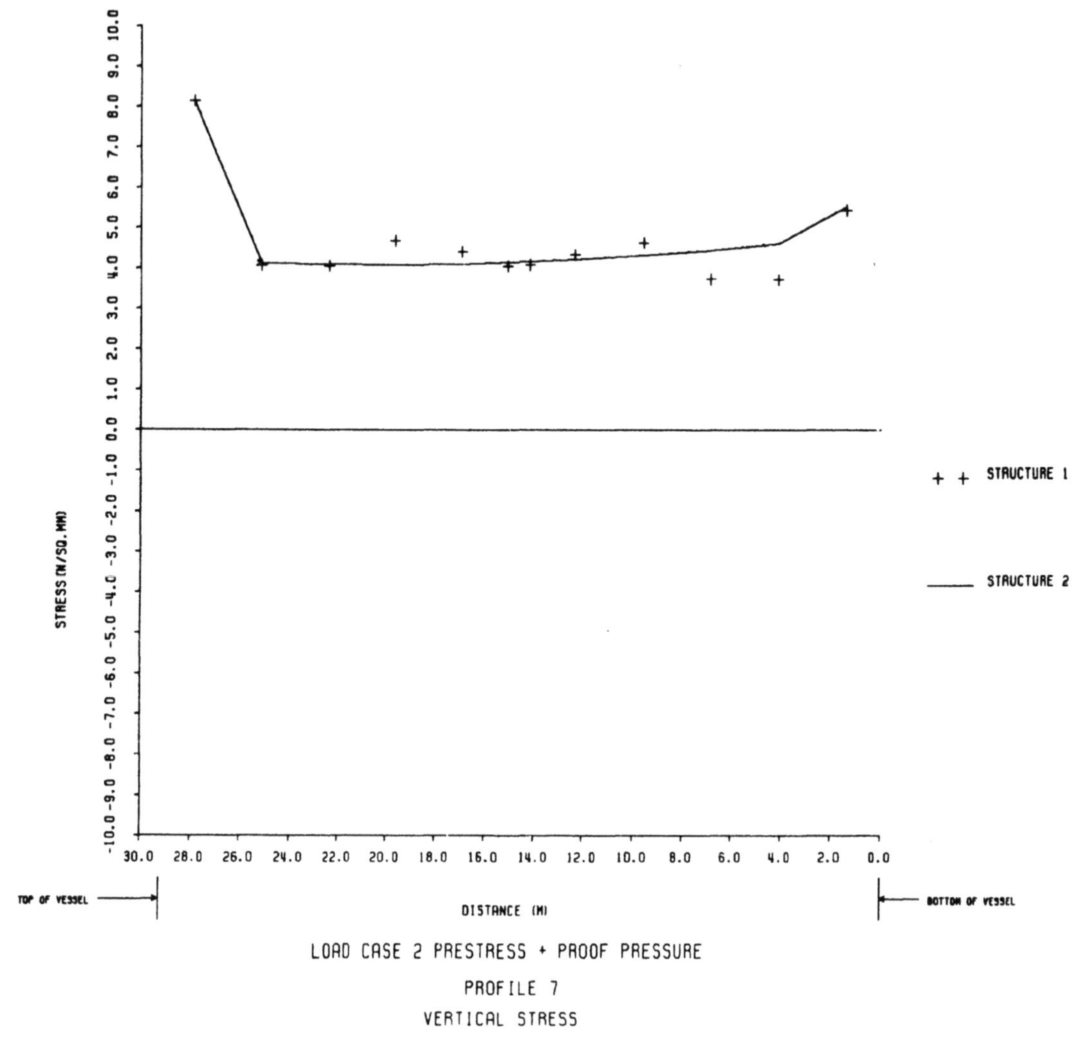

FIG. 33 PLANES OF WEAKNESS ANALYSIS
TYPICAL COMPARISON OF STRESS PROFILES FOR
STRUCTURES TYPE 1 AND 2

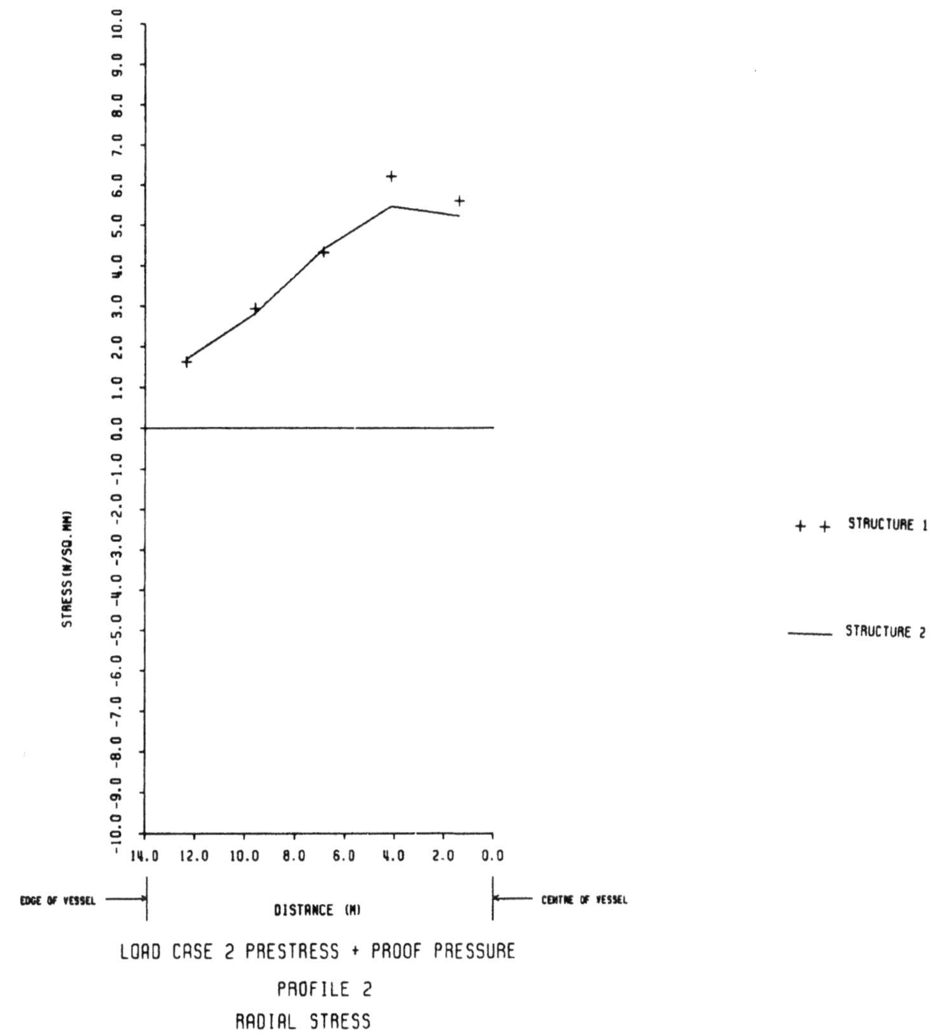

FIG. 34 PLANES OF WEAKNESS ANALYSIS
TYPICAL COMPARISON OF STRESS PROFILES FOR
STRUCTURES TYPE 1 AND 2

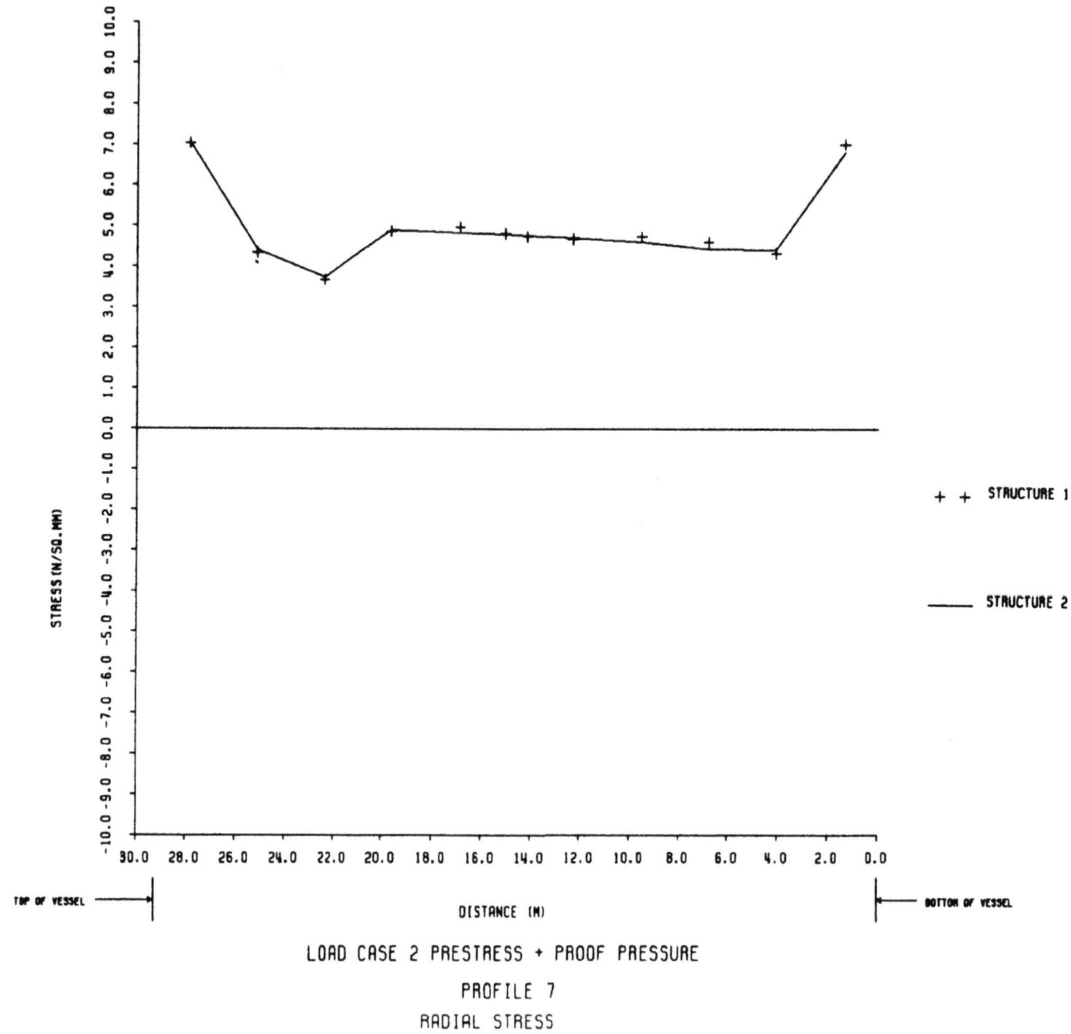

FIG. 35 PLANES OF WEAKNESS ANALYSIS
TYPICAL COMPARISON OF STRESS PROFILES FOR
STRUCTURES TYPE 1 AND 2

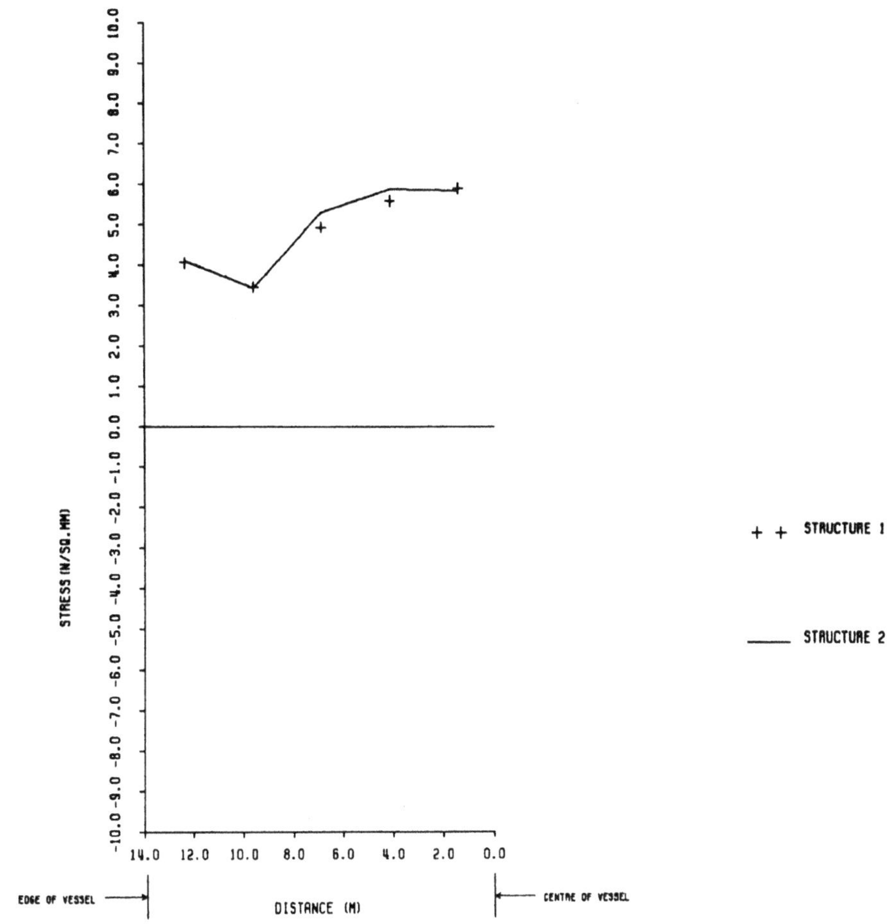

LOAD CASE 2 PRESTRESS + PROOF PRESSURE

PROFILE 2

HOOP STRESS

FIG. 36 PLANES OF WEAKNESS ANALYSIS
TYPICAL COMPARISON OF STRESS PROFILES FOR
STRUCTURES TYPE 1 AND 2

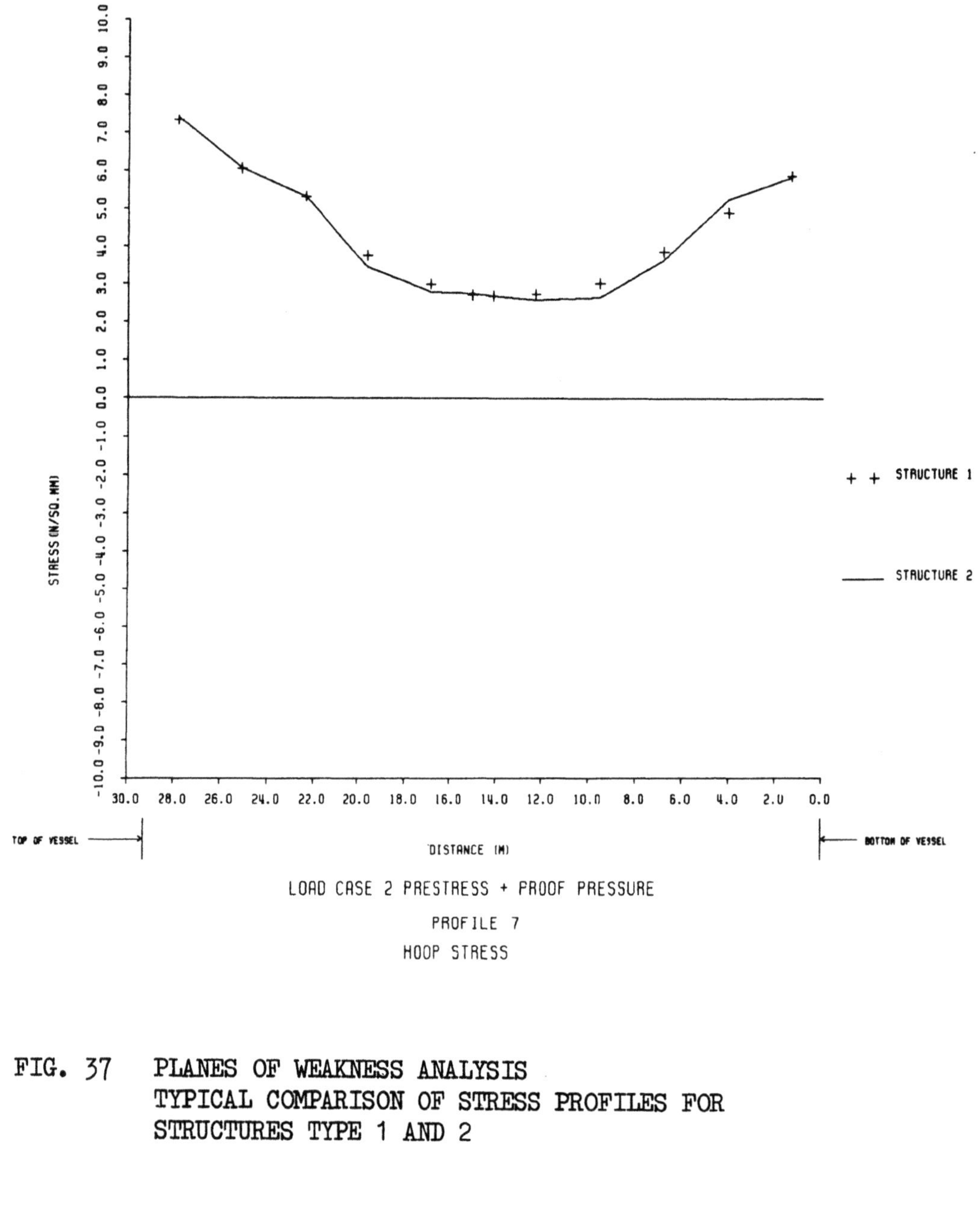

FIG. 37 PLANES OF WEAKNESS ANALYSIS
TYPICAL COMPARISON OF STRESS PROFILES FOR
STRUCTURES TYPE 1 AND 2

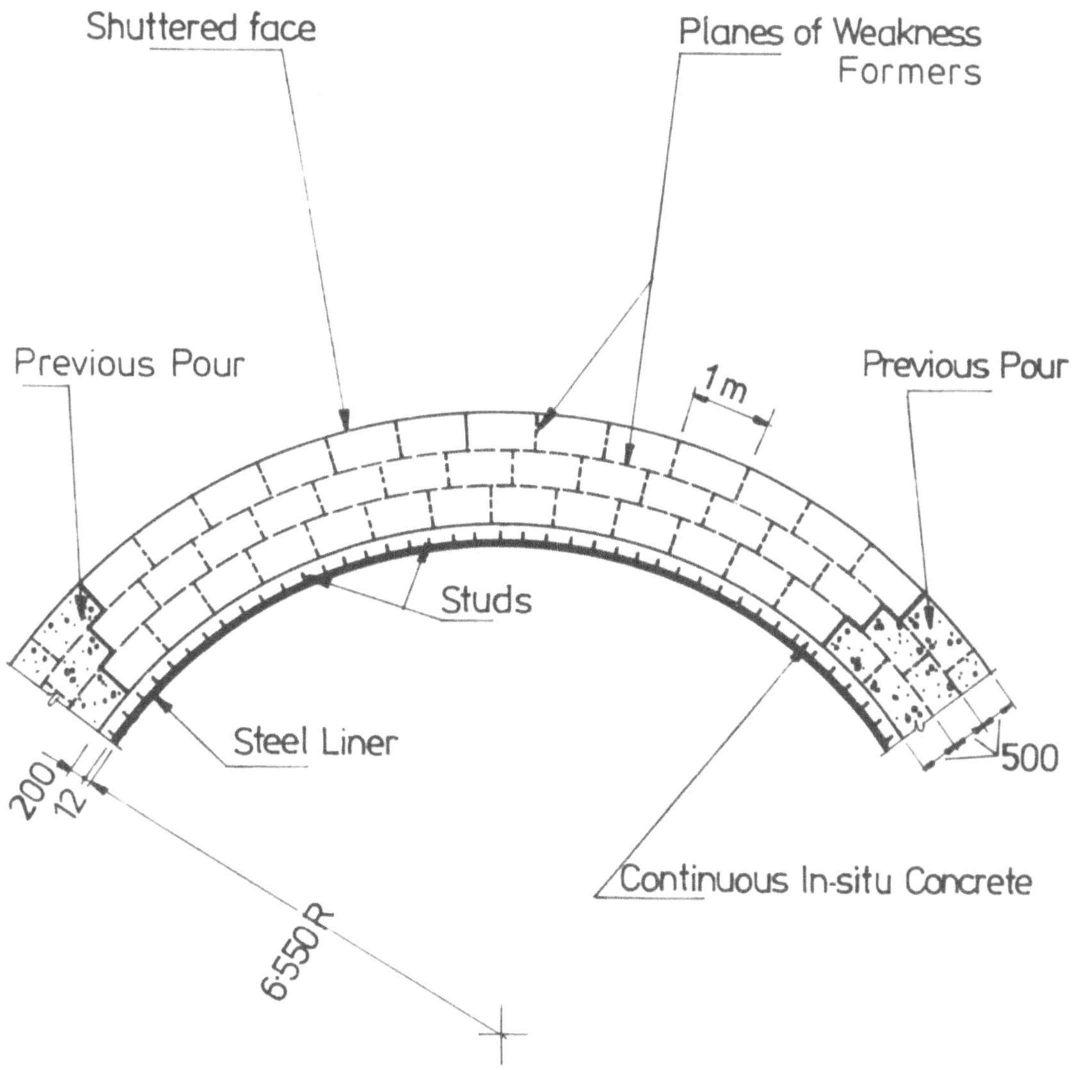

Fig.38. IN-SITU CONSTRUCTION FORMATION OF
PLANES OF WEAKNESS WITHIN
A NORMAL SIZED POUR

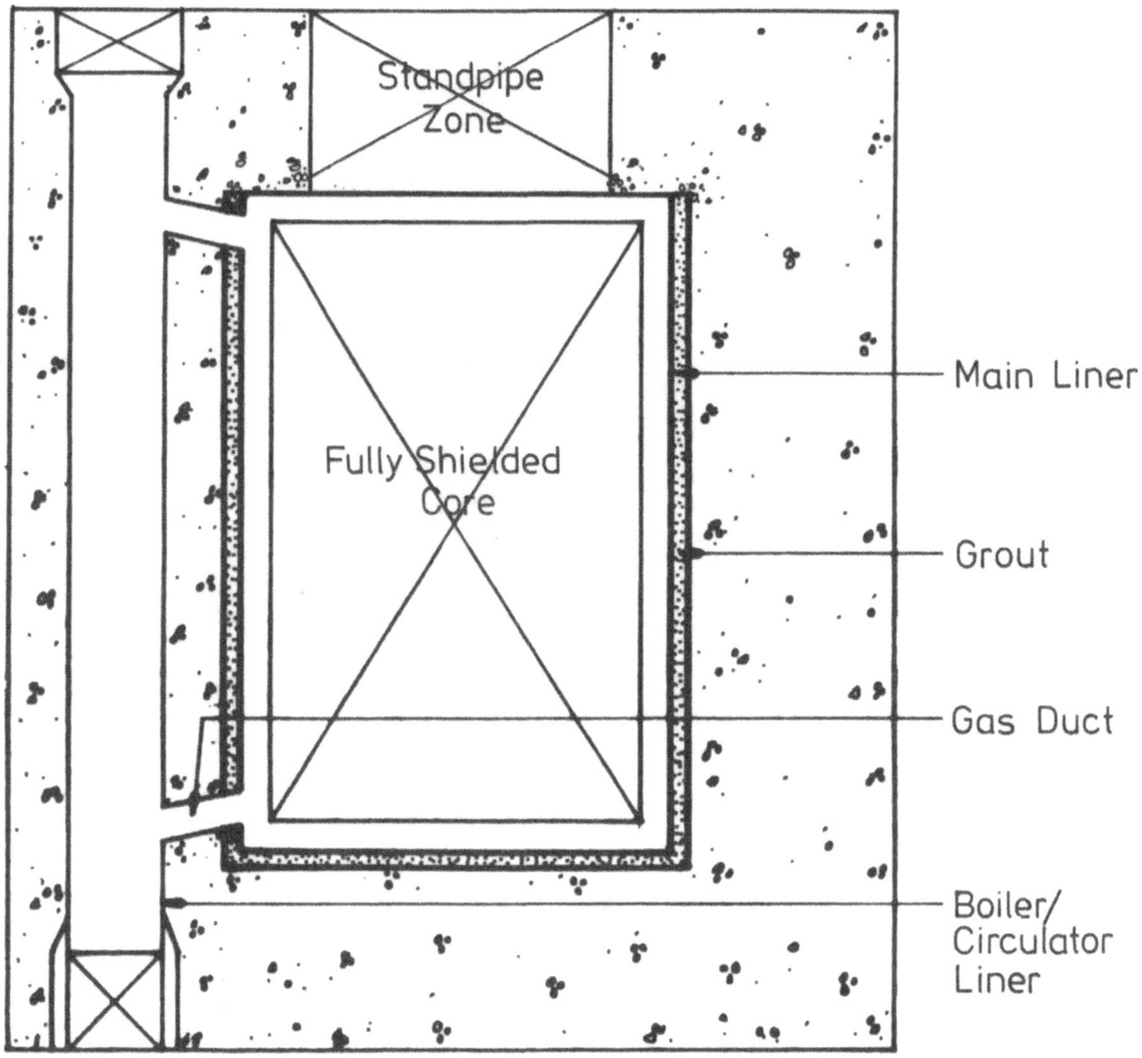

Fig.39. OUTLINE PROPOSAL FOR GROUTED LINER

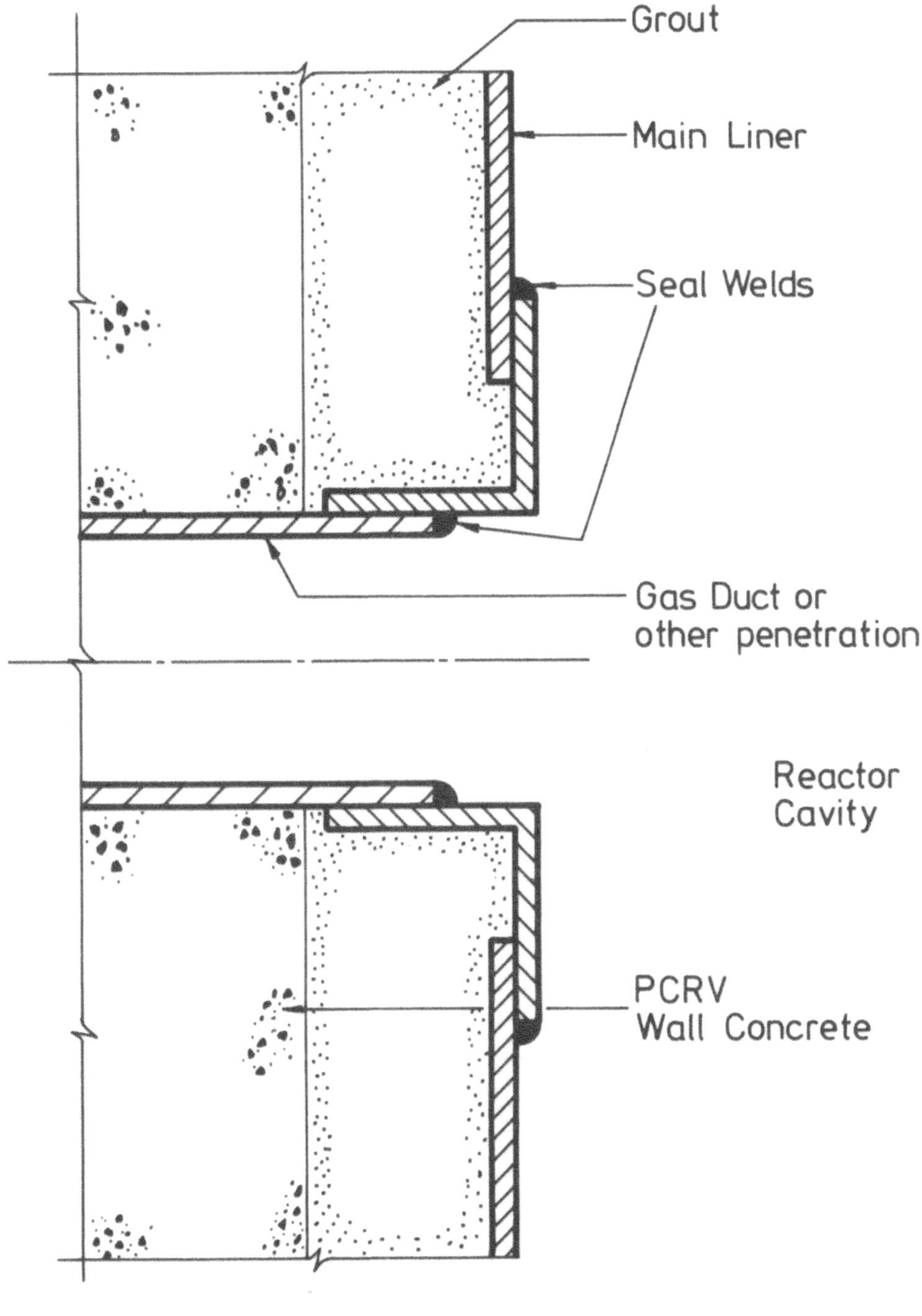

Fig.40. PROPOSED DETAIL OF JUNCTION BETWEEN GROUTED MAIN LINER AND GAS DUCT

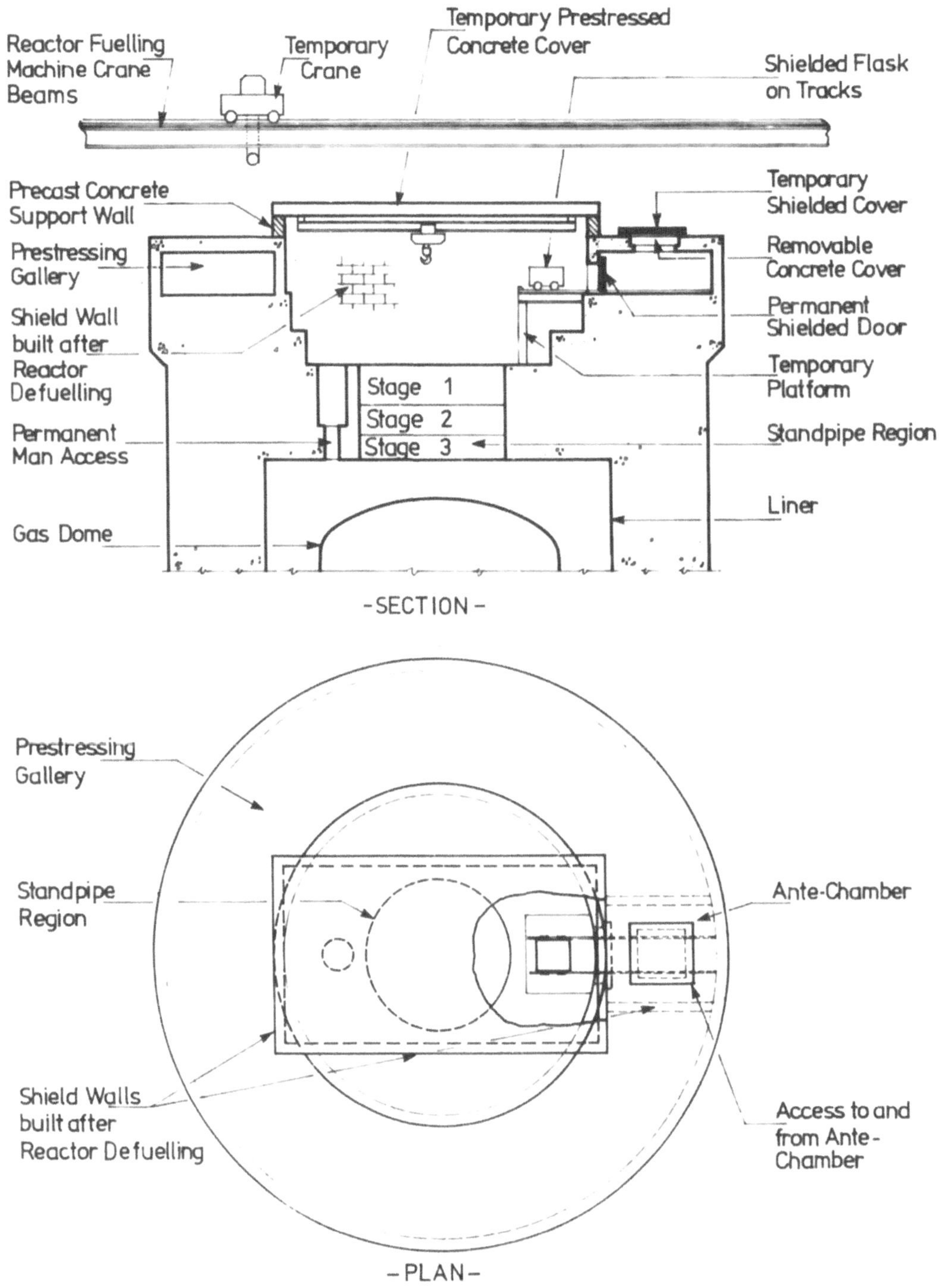

Fig 41 OUTLINE SCHEME FOR REMOVAL OF STANDPIPE REGION TO GIVE ACCESS FOR REMOVAL OF REACTOR CORE AND INTERNALS

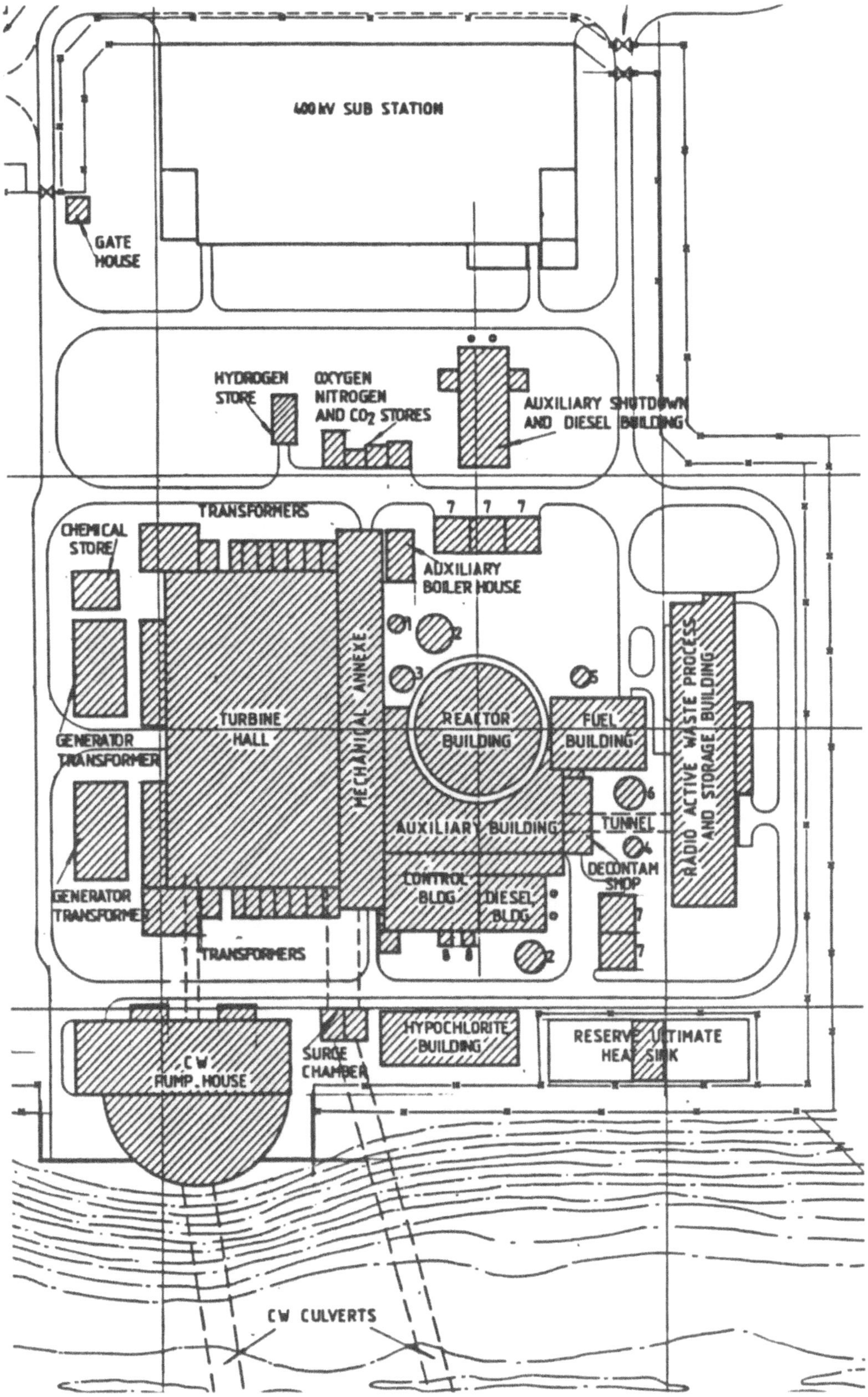

Fig.42. SITE PLAN – TYPICAL PWR STATION

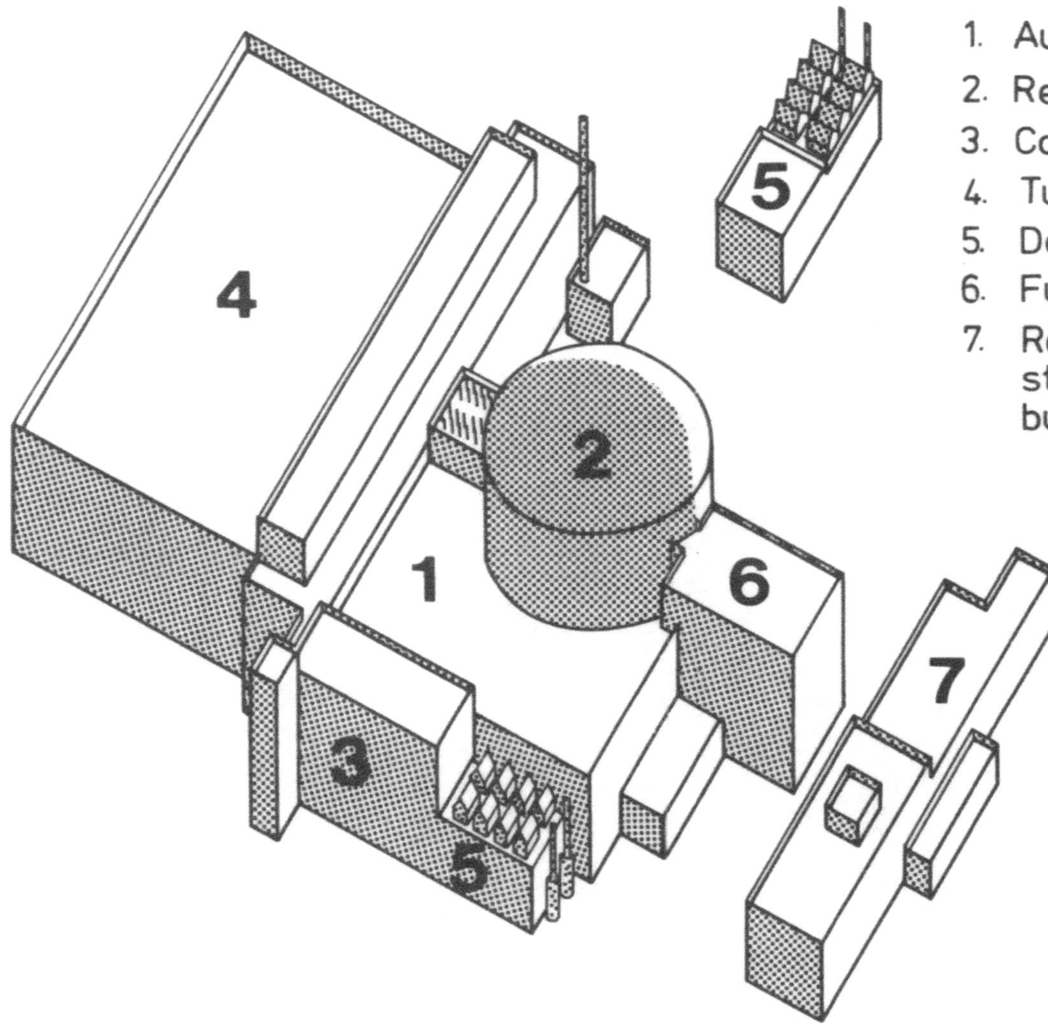

1. Auxilliary building.
2. Reactor building.
3. Control building.
4. Turbine building.
5. Deisel building.
6. Fuel building.
7. Radioactive waste storage & processing building.

Fig 43 MAJOR STRUCTURES OF TYPICAL P.W.R. STATION

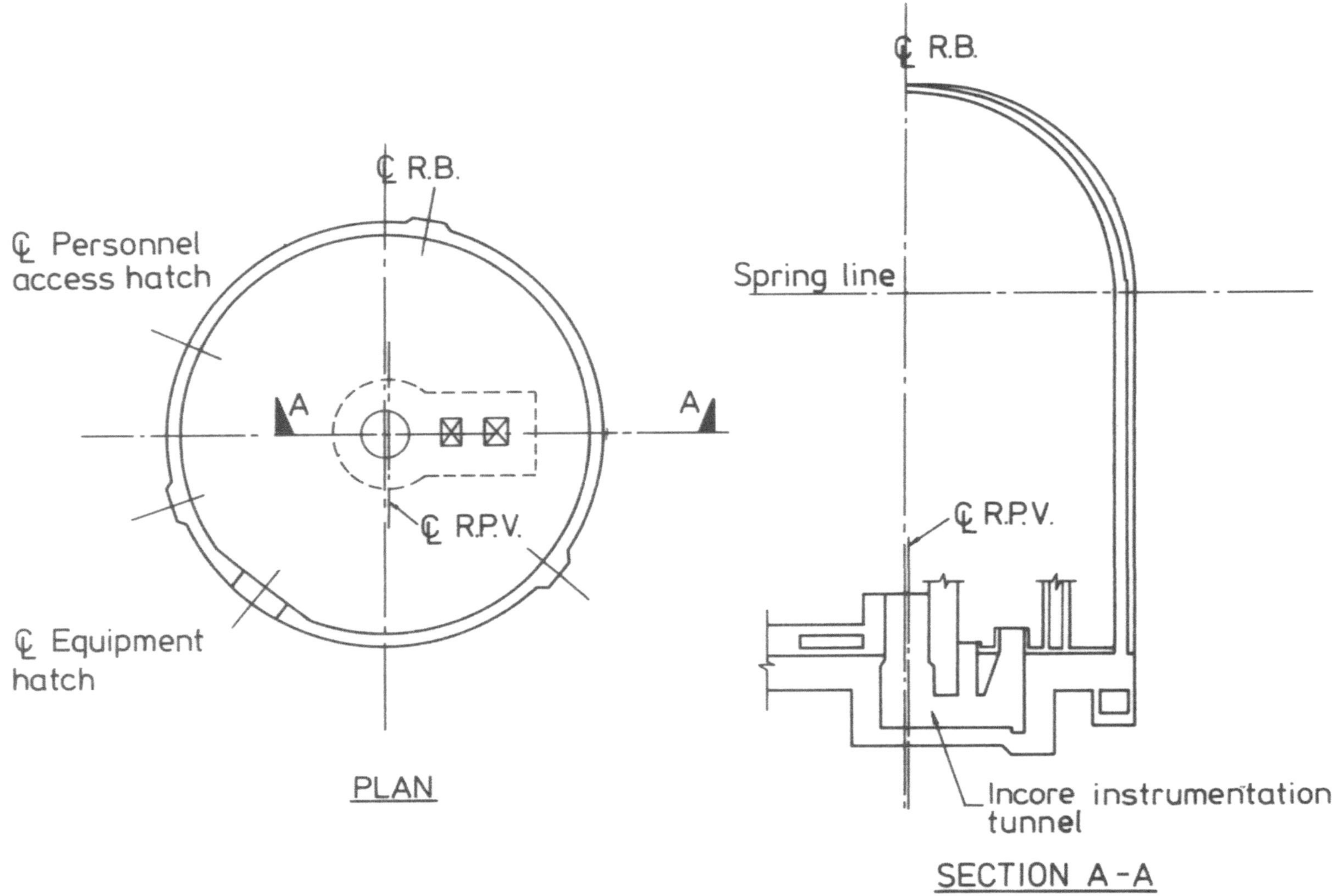

Fig.44. REACTOR BUILDING OUTLINE OF TYPICAL PWR STATION

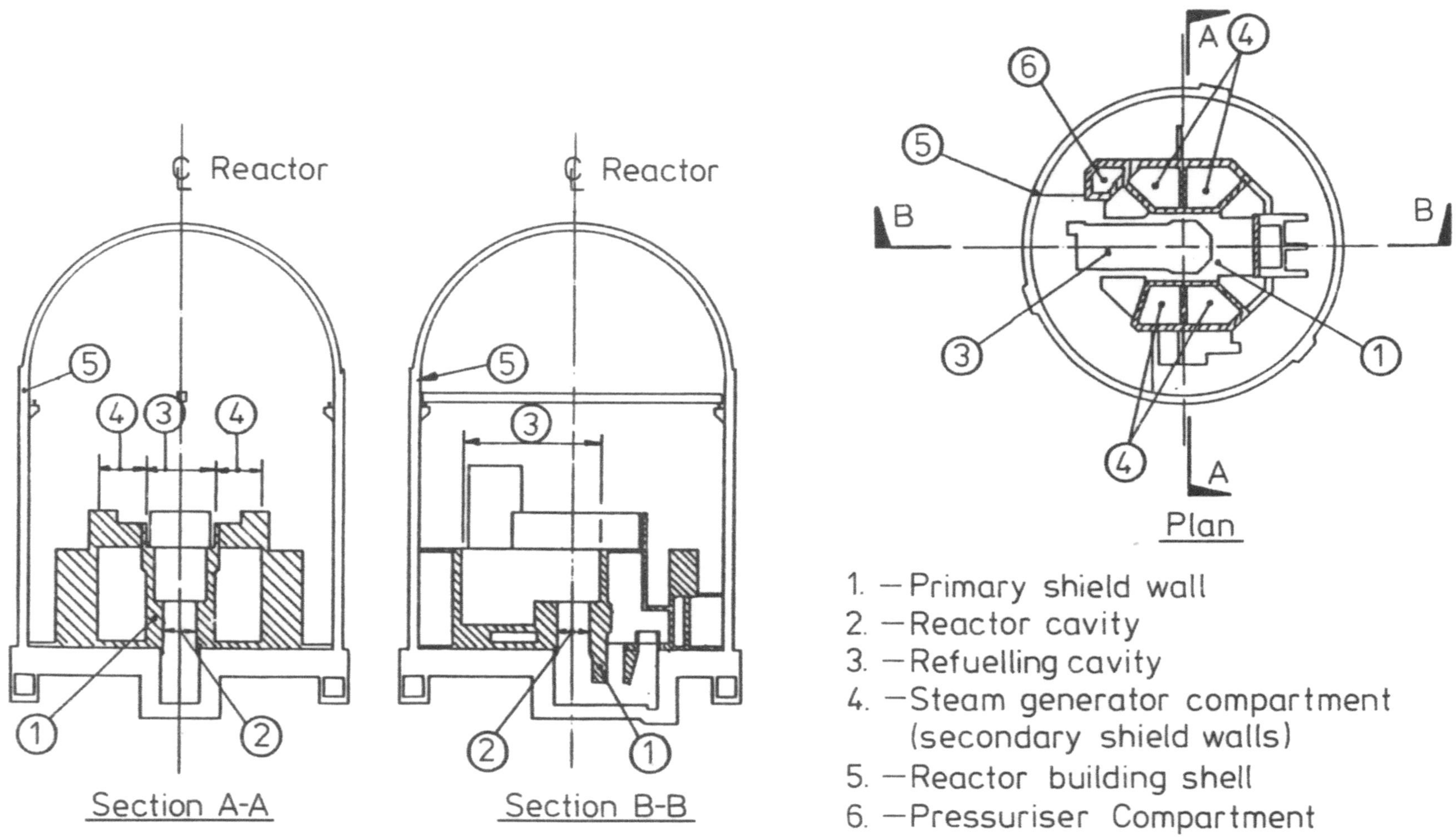

Fig 45 REACTOR BUILDING STRUCTURES OF TYPICAL PWR STATION

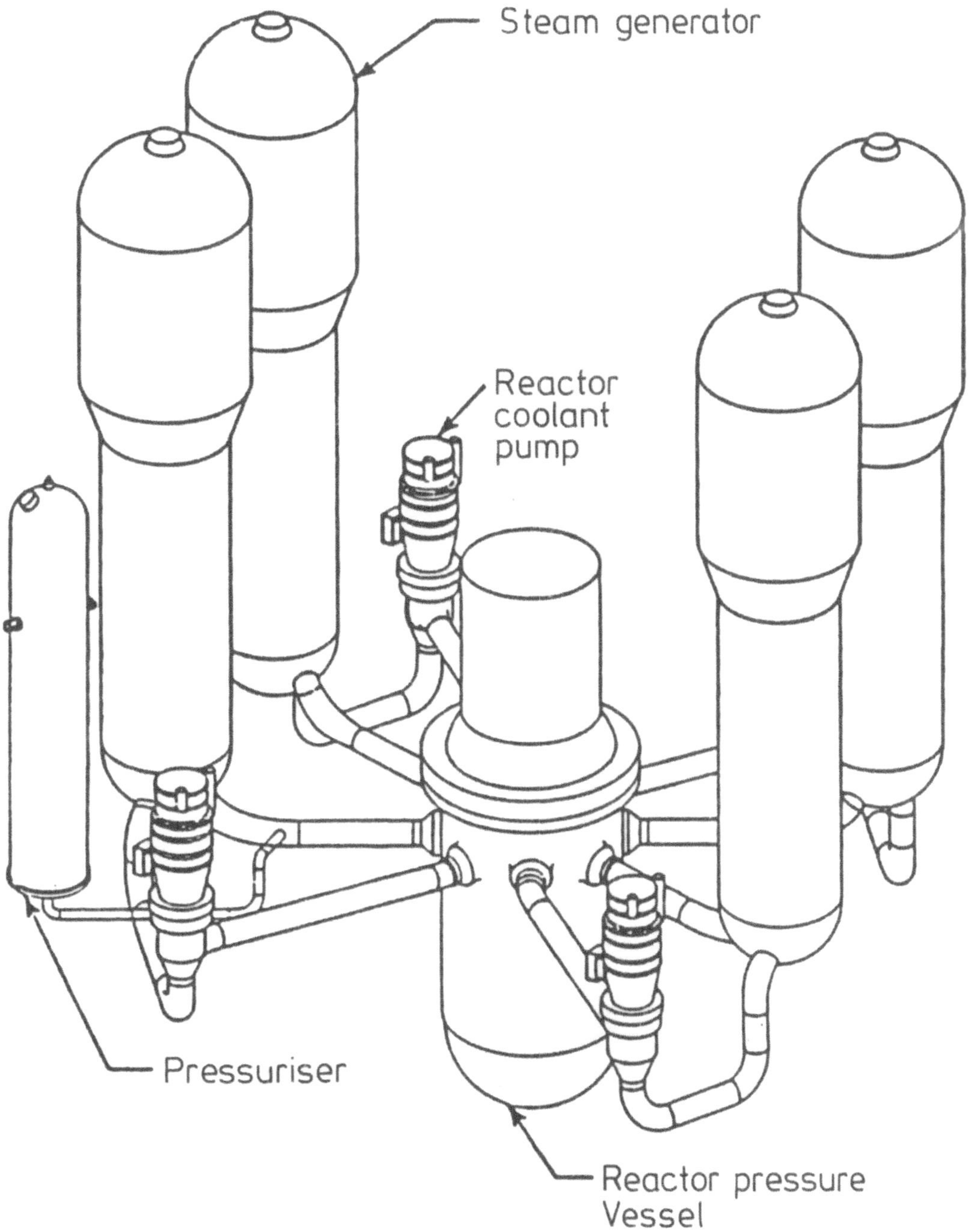

Fig.46. NUCLEAR STEAM SUPPLY SYSTEM
OF TYPICAL PWR.

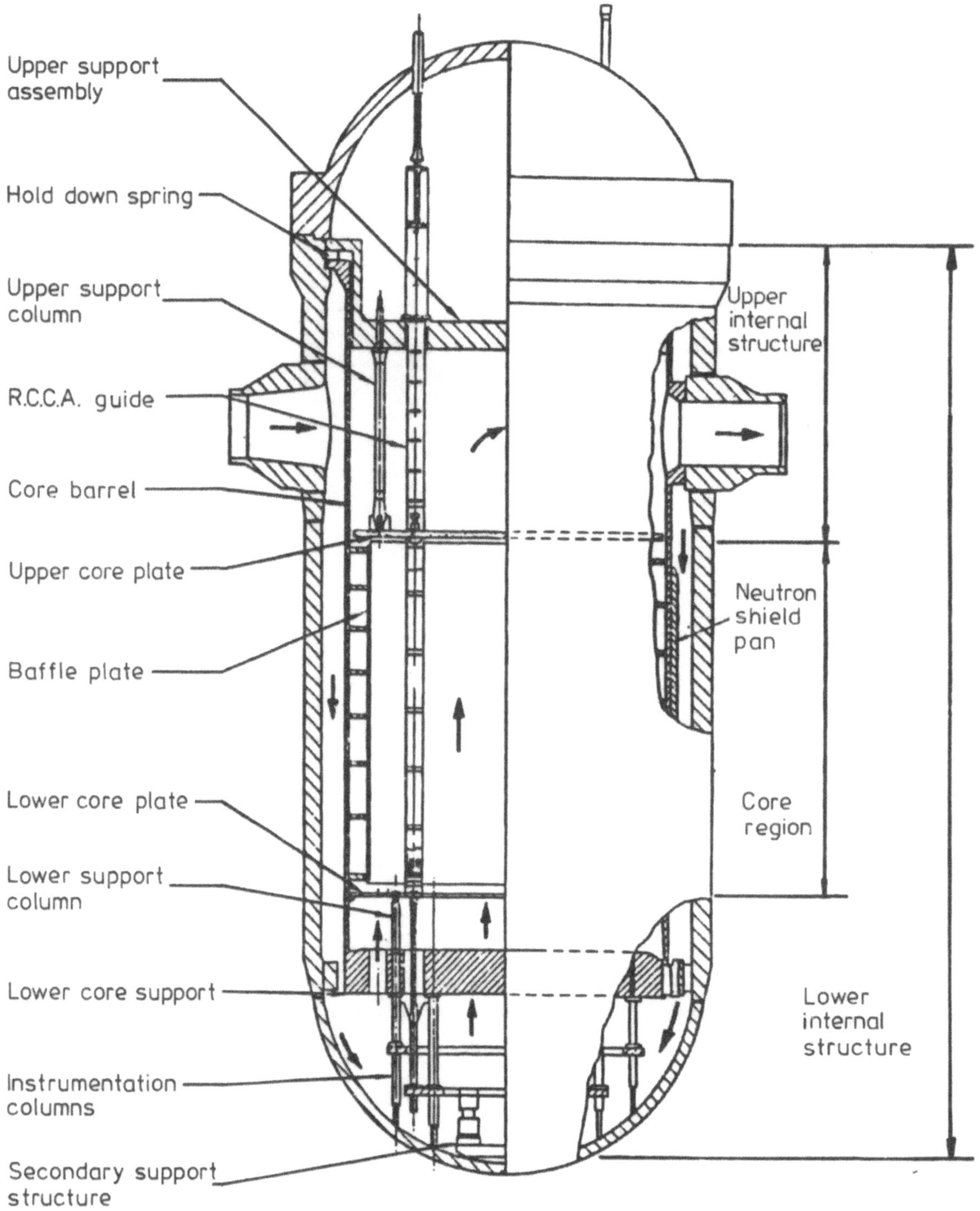

Fig.47. SECTION THROUGH REACTOR PRESSURE VESSEL OF TYPICAL PWR

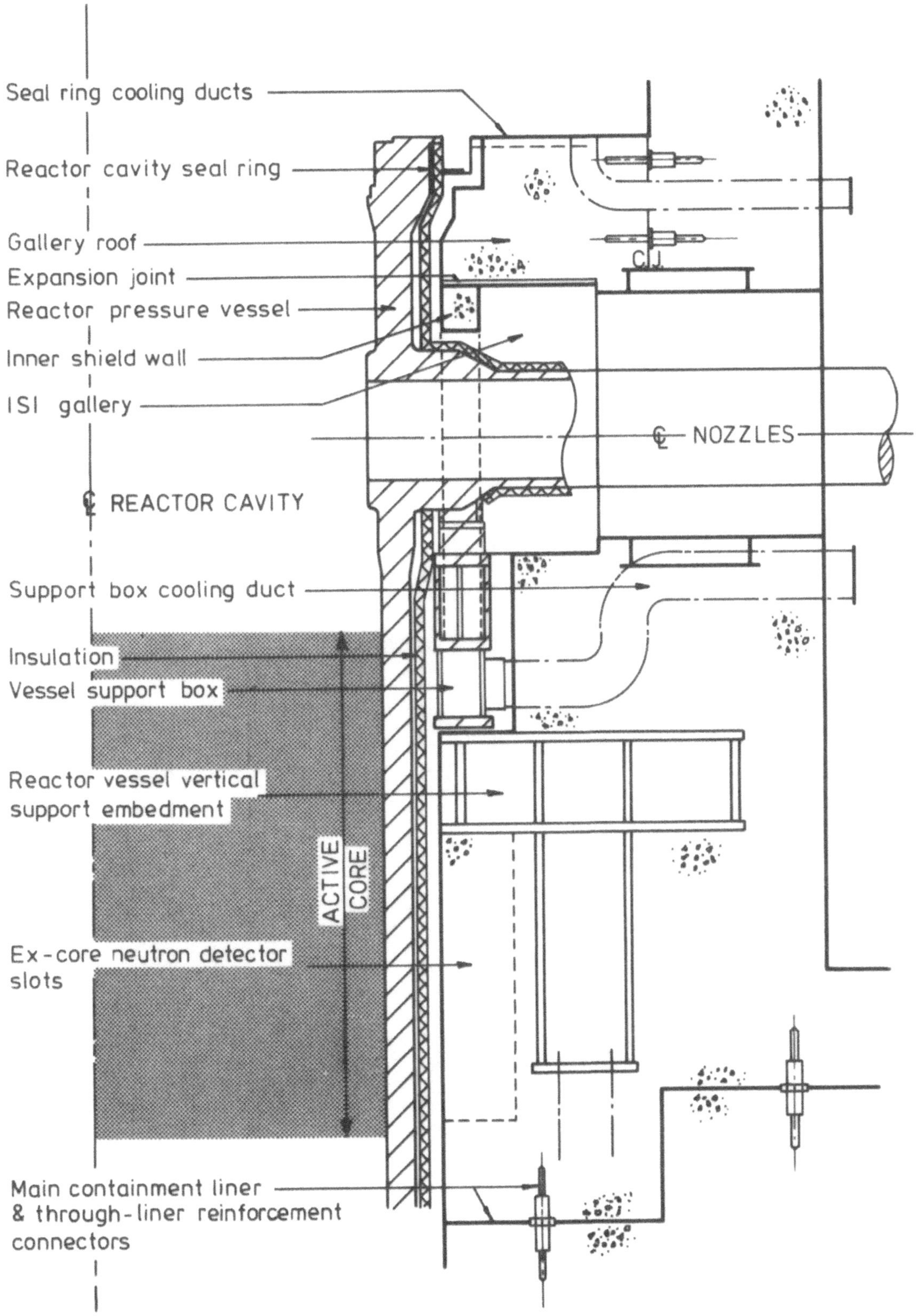

Fig 48 SECTION THROUGH LOOP PIPE PENETRATION OF TYPICAL PWR PRESSURE VESSEL

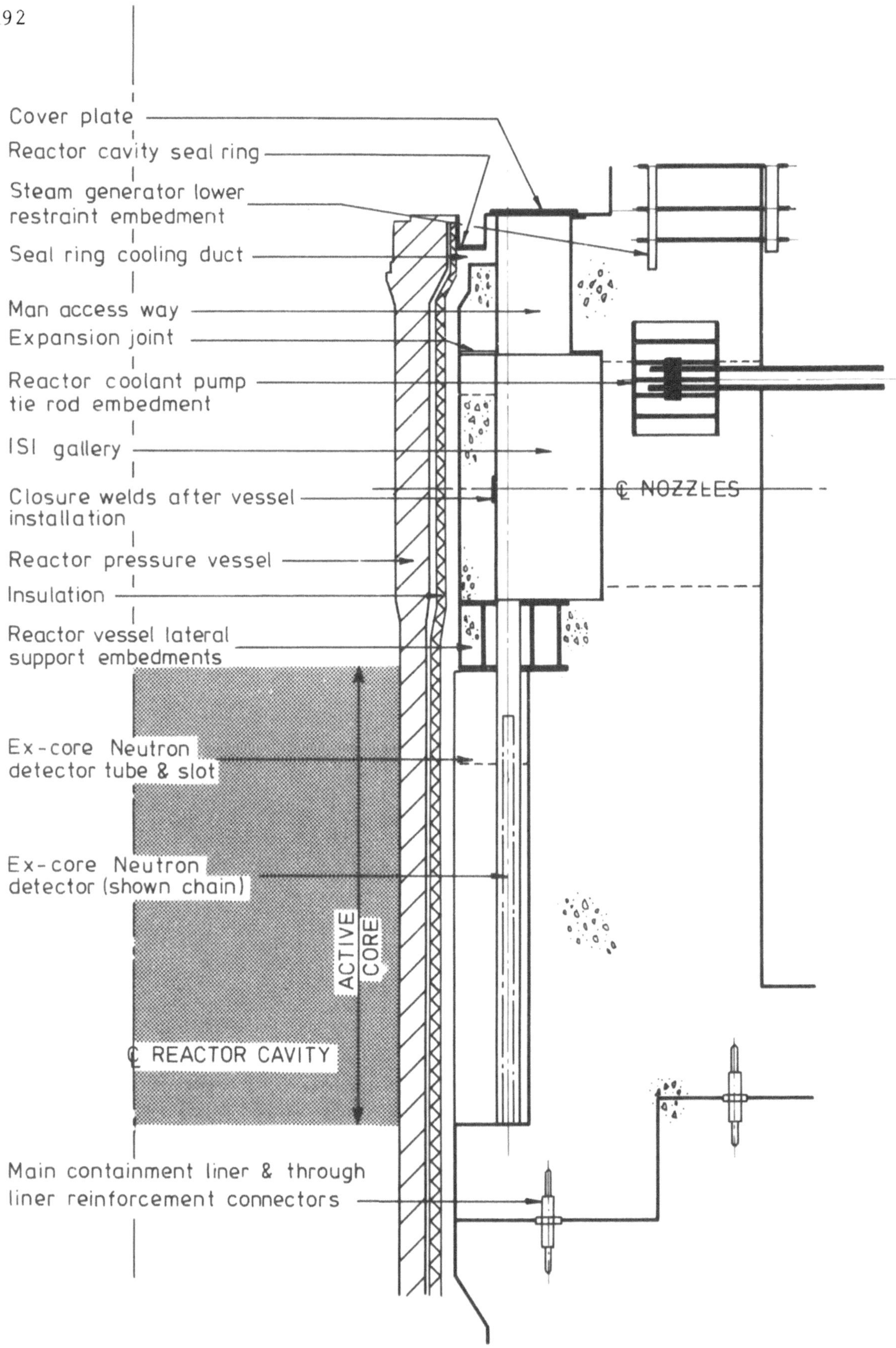

Fig. 49. SECTION THROUGH DETECTOR SLOT OF TYPICAL PWR PRESSURE VESSEL

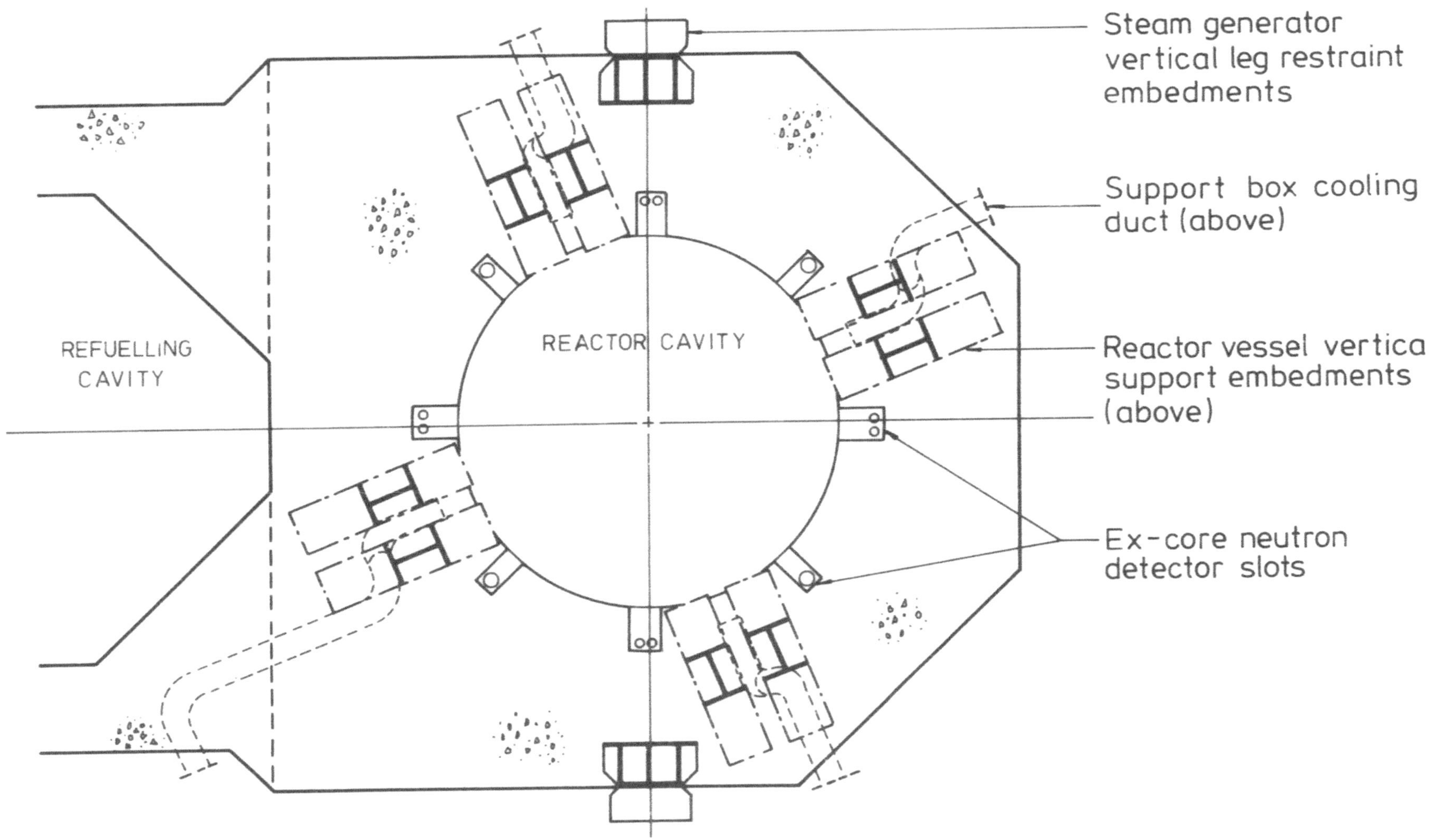

Fig.50. PLAN SHOWING DETECTOR SLOTS OF TYPICAL PWR PRESSURE VESSEL

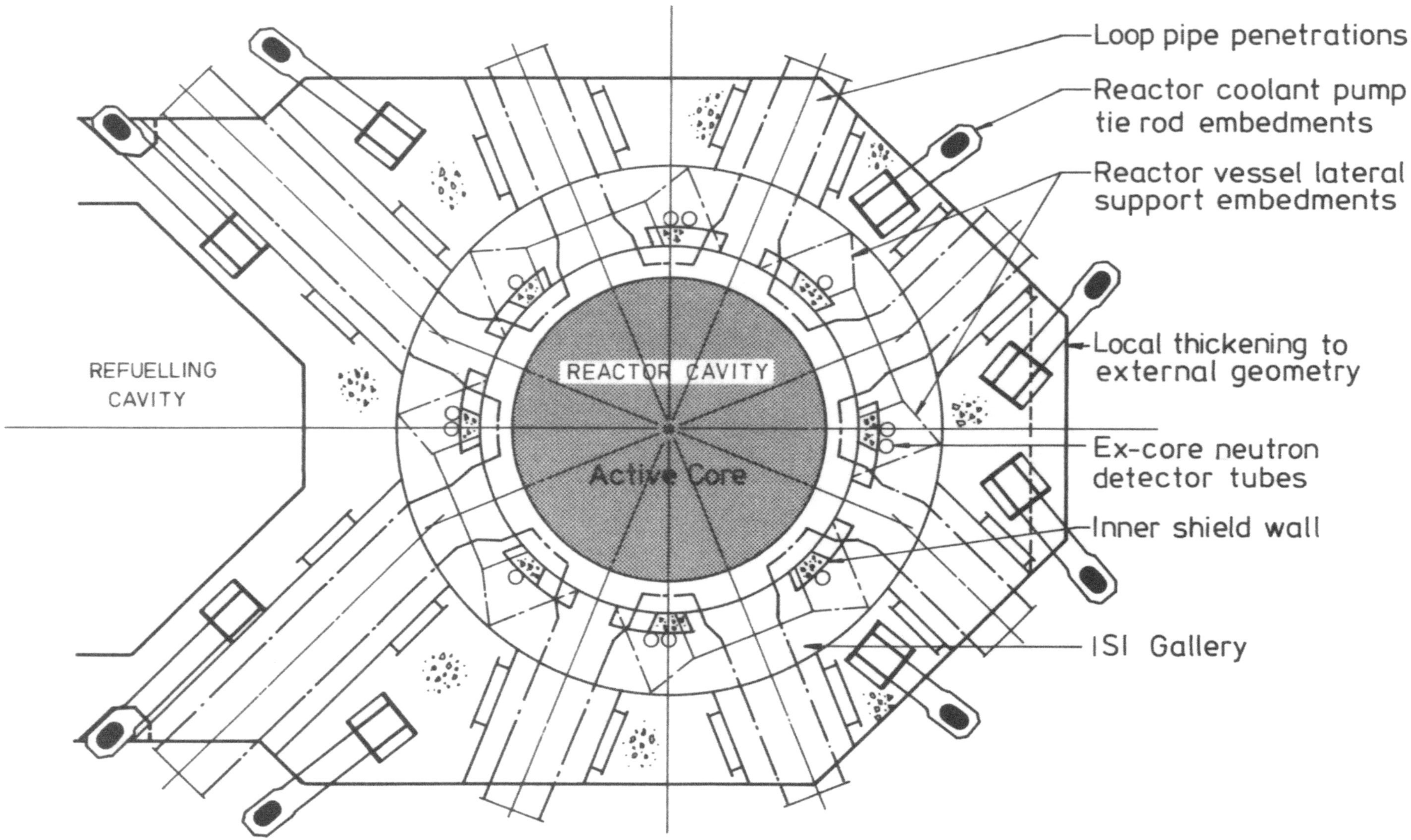

Fig.51. PLAN ON ISI GALLERY OF TYPICAL PWR PRESSURE VESSEL

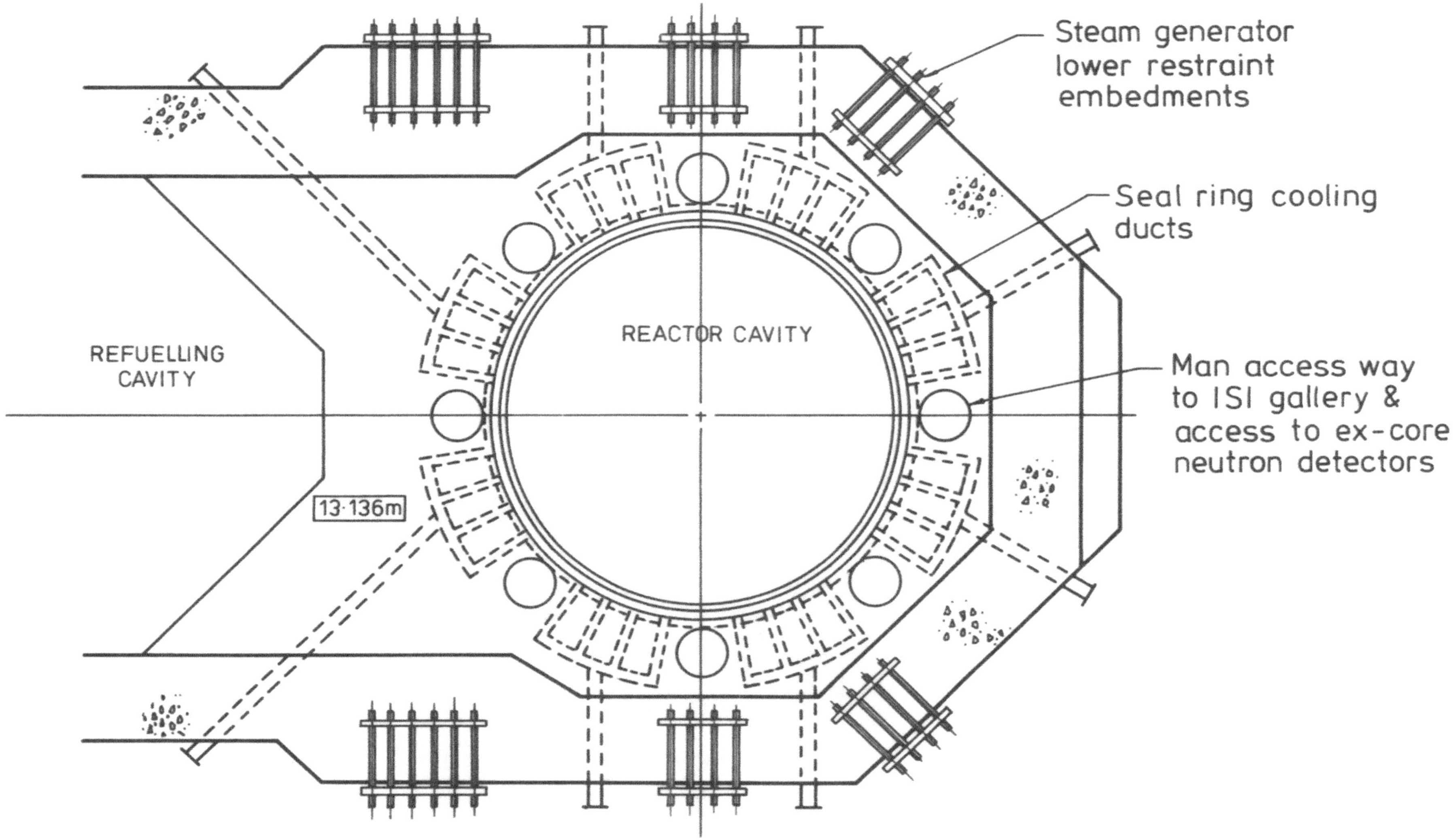

Fig.52 PLAN ON FUELLING POOL OF TYPICAL PWR PRESSURE VESSEL

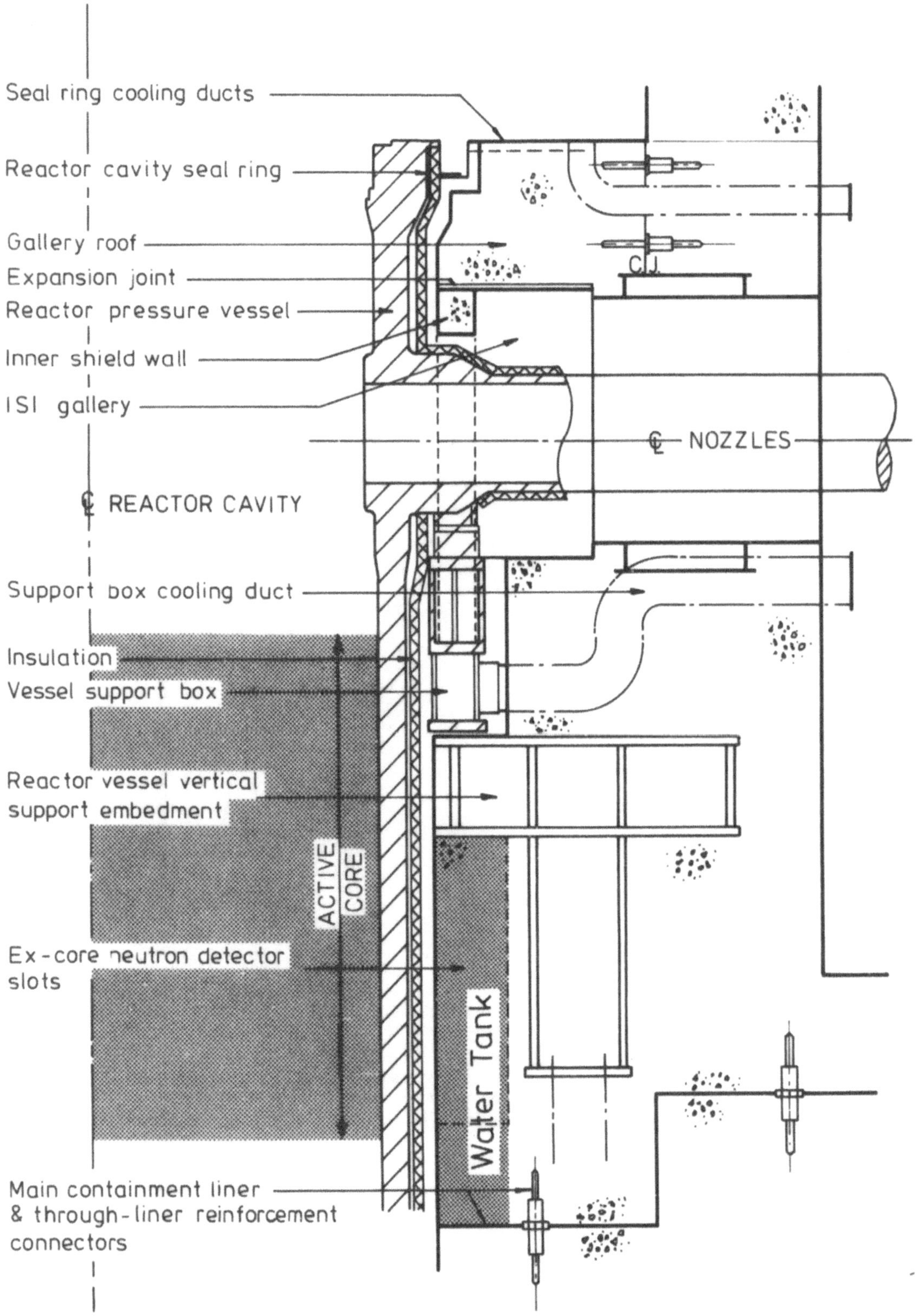

Fig.53. SECTION SHOWING POSSIBLE LOCATION OF WATER TANKS IN PWR PRIMARY SHIELD WALL

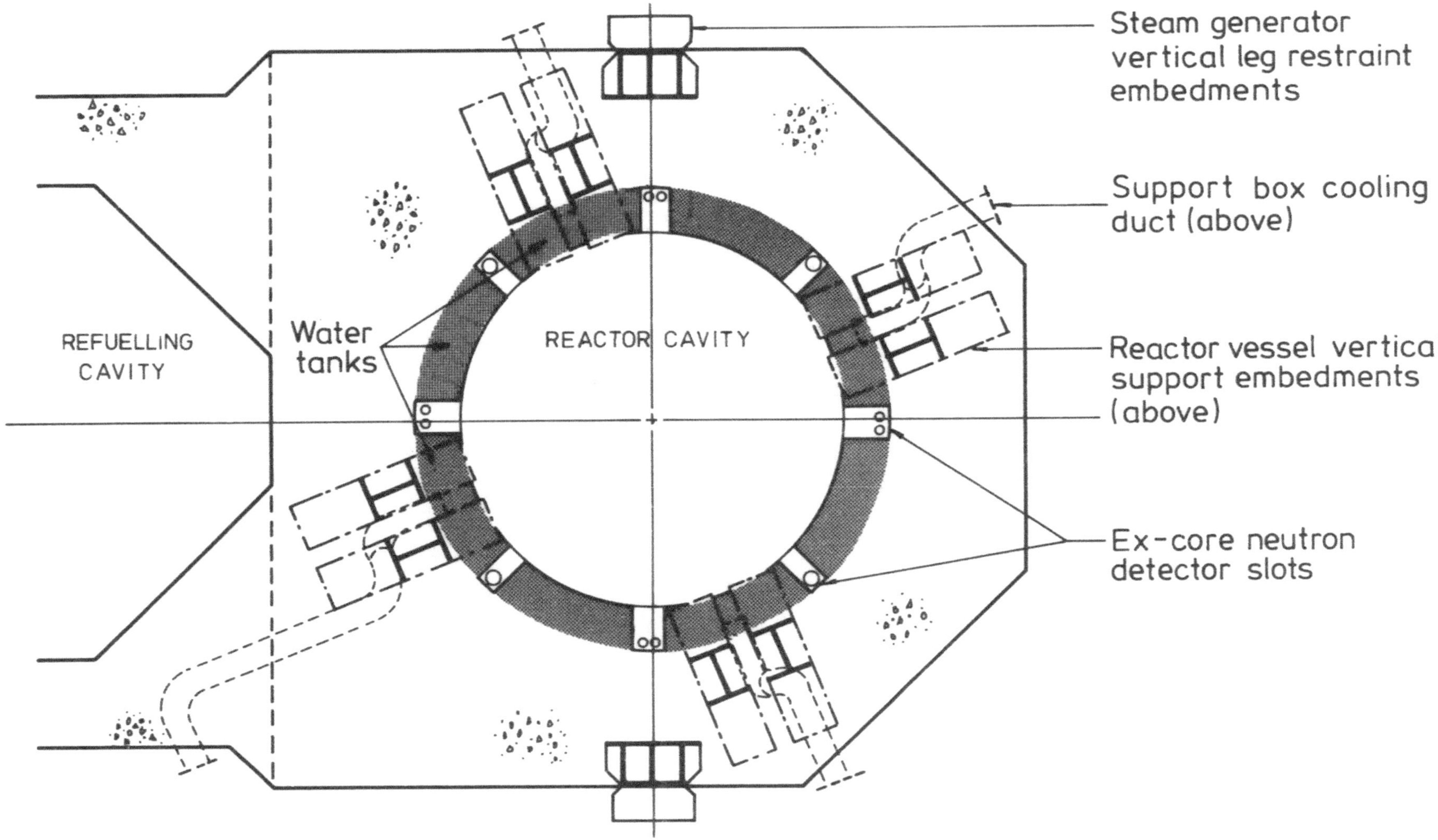

Fig 54 PLAN SHOWING POSSIBLE LOCATION OF WATER TANKS IN PWR PRIMARY SHIELD WALL

APPENDIX A

SUPPLEMENTARY INFORMATION

This appendix contains additional information and discussion relating to a number of ideas mentioned but not fully described in the main report.

(a) Provision of preformed holes for the placement of explosive charges.

The use of explosives for the controlled demolition of concrete biological shields and pressure vessels is being investigated by TWC under Project 3 of the CEC programme on nuclear plant decommissioning. This work has shown that small explosive charges placed in holes formed at right angles to the inside face of a massive concrete structure can be used to strip off concrete layers from the face. Each concrete layer stripping operation is carried out using predetermined charge spacing and depth of burial designed to give optimum results with respect to safety, volume of material excavated and economy. The layer by layer stripping operation is continued until the required thickness of concrete has been cratered off, or excavated. The indications are that a one metre thickness could be fairly satisfactorily removed in a series of four layer stripping operations. If a mat of reinforcement is present close to the face of the concrete, the effect of the first layer round of firings will be to blow the cover off and, depending on the reinforcement detail, break up some concrete behind the reinforcement. The reinforcement mat so exposed can be cropped or cut away to release any debris held behind it. Once the reinforcement mat has been removed the stripping process proceeds as normal.

No tests have been carried out to assess the effect of a liner bonded to the concrete by studs. It may be that, if the liner is cut through into discrete plates consistent with the expected pattern of explosive cratering, the concrete can be blown off in discrete pieces with the liner still attached to it. This would, however, have to be investigated by a programme of testing.

The layout of boreholes required to strip a given thickness, say one metre, of concrete could be predetermined at the design stage. This being so it would be possible to form holes in the concrete during construction of the vessel subject to the maintenance of adequate shielding. This would permit the placement of charges during decommissioning without having to bore or cut holes first. Boring or cutting such holes would produce a certain amount of activated waste which would have to be suitably disposed of. It would also be a time consuming process, increasing radiation doses received by decommissioning workers. A preferable approach would be to use shaped charges which are faster and produce negligible active waste.

There are, however, two main problems with preformed holes:-

(1) Shine paths for radiation.
(2) Integrity of the liner.

The concentration of holes will be such that shine paths for radiation could be a problem, forming 'hot spots' of activation outside of the designated activated zone. If unacceptable, this could be overcome if the holes were plugged with a material with good neutron shielding properties. Assuming that the concrete is cast up against the liner, the hole filler would have to be a solid object, such as graphite or lead, fixed in position before concrete placing, suitably debonded from the concrete. The size of the holes would be limited by the ability of the liner to span the hole, and whether the material in the hole could support the liner. In carrying out the ultimate load analysis for the Hartlepool/Heysham I vessels, crack width was limited to 32mm for this reason. It is not thought likely that the density of holes would cause major problems due to stress concentrations but this would have to be checked following design of a suitable layout of charges.

(b) <u>Provision of preformed holes for the placement of a chemical demolition agent.</u>

The use of a non-explosive chemical demolition agent to split up concrete overcomes problems due to noise, vibration, dust, flying debris, etc. associated with the use of explosives. If mixed with water and poured into cylindrical holes formed in the concrete, "BRISTAR" expands, generating cracks in the surrounding concrete. This product has been successfully used for breaking up rock and massive unreinforced concrete structures. For reinforced concrete, a close hole spacing of about 200mm to 300mm is required, hole diameters being 40mm to 50mm. Provision of preformed holes subject to provisos mentioned in item (a) above would aid the use of such an agent, but the number of holes would be such that stress concentrations could become quite significant.

(c) <u>Provision of preformed cavities for the insertion of mechanical burster.</u>

There are two types of hydraulic burster, namely wedge type and piston type, and both split the concrete by expanding outwards against the circumference of a suitable hole into which they have been placed. The spacing of holes, their position, depth and angle govern the success of bursting. Generally speaking, the hole for the insertion of the burster should be less than 500mm from a free face. Alternatively, a number of bursters can be used simultaneously, in holes about 600mm apart in a line about 600mm from the free face. Hole diameter depends on the type of burster being used. Provided a free working face could first be established, hydraulic bursters could provide a controlled means of splitting up activated concrete for removal. The provision of preformed holes, subject again to the provisos mentioned in item (a) above would save the time consuming and expensive job of cutting or boring holes.

(d) Improvement in the neutron shielding properties of concrete to reduce the volume of activated concrete and surface dose rates.

Activation of the concrete in a PCRV is caused by production of radioactive isotopes due to capture of neutrons by the nuclei of certain elements present in the concrete. Before capture, fast neutrons must be slowed down to thermal levels. Hydrogen is very effective at slowing down neutrons over a wide range of energies. By increasing the amount of hydrogen in concrete, the neutron attenuating properties of the concrete can be improved and two methods of increasing the proportion of hydrogen in hardened concrete are:

(1) Increasing the amount of free and chemically bound water in the hardened cement paste.
(2) Using hydrous aggregates.

Method (1) could involve increasing the fineness of the cement and increasing the cement and water content of the mix combined with stringent curing procedures during construction. Migration of free water through concrete is very slow, even under PCRV operating temperatures. Naturally occurring aggregates with a high hydrogen content include limonite, geophite and serpentine.

The adoption of any of these measures would have to be preceded by trials to establish the effects on workability, heat of hydration during construction, creep and shrinkage properties and long term strength and durability of the resulting concrete.

Boron is very effective in capturing thermal neutrons and does not produce gamma-emitting isotopes. The addition of boron to concrete will reduce dose rates received by decommissioning workers from harmful gamma radiation. Some of the more readily available boron containing additives are soluble in water and have a deleterious effect on the concrete setting process. As a result, borated concretes used in practice in reactor construction in the U.S.A. and Japan have typically had boron contents of less than 1% by weight. However, Japanese researchers have succeeded in producing concrete mixes with a boron content of up to 6.4% using non-soluble boron frits. The presence of these appears to retard slightly the early gain of strength, but 28 day strengths similar to or even greater than comparable mixes without boron are achieved, and all of the trial mixes reported had 28 day strengths of greater than the 40 N/mm^2 commonly specified for PCRVs. Incorporation of boron in the PCRV concrete may also affect reactor operating characteristics.

(e) Use of an alternative material to mild steel for the liner.

Discussions with liner design engineers have revealed that alternative materials to steel for liners have been investigated in the past. One of the important properties required by material used on heavy civil engineering construction sites is durability against damage from both normal site operations and misuse of equipment such as dropped scaffold tubes, reinforcing rods and the like. Recent tests on high density polyethylene sheet have shown that this material can be easily penetrated by dropped objects and, although easily repaired, the integrity of the liner would depend on all such damage being identified before the liner is built in or covered up. Plastic materials used in industry for containment liners have good chemical resistance and elasticity properties but are not able to withstand the high temperatures normally found in a reactor. The long term effect of radiation on some plastics, with resulting embrittlement, is also a problem to be taken into account.

Stainless steel is used in nuclear installations as an impermeable containment membrane but this material is normally used where corrosive conditions prevail and where ease of decontamination is required. Compared with mild steel, stainless steel is considerably more expensive and exhibits a higher level of harmful radioactive isotopes, such as nickel, when subjected to radiation. All enquiries to date have failed to identify a better material than mild steel for the liner.

GPSR Compliance

The European Union's (EU) General Product Safety Regulation (GPSR) is a set of rules that requires consumer products to be safe and our obligations to ensure this.

If you have any concerns about our products, you can contact us on ProductSafety@springernature.com

In case Publisher is established outside the EU, the EU authorized representative is:

Springer Nature Customer Service Center GmbH
Europaplatz 3
69115 Heidelberg, Germany

Batch number: 10615097

Printed by Printforce, the Netherlands